CAMBRIDGE LIBRARY COLLECTION

Books of enduring scholarly value

Polar Exploration

This series includes accounts, by eye-witnesses and contemporaries, of early expeditions to the Arctic and the Antarctic. Huge resources were invested in such endeavours, particularly the search for the North-West Passage, which, if successful, promised enormous strategic and commercial rewards. Cartographers and scientists travelled with many of the expeditions, and their work made important contributions to earth sciences, climatology, botany and zoology. They also brought back anthropological information about the indigenous peoples of the Arctic region and the southern fringes of the American continent. The series further includes dramatic and poignant accounts of the harsh realities of working in extreme conditions and utter isolation in bygone centuries.

The Atlantic

In 1871 the British government agreed to support an expedition to collect physical and chemical data and biological specimens from the world's oceans. Led by Charles Wyville Thomson (1830–82), the expedition used H.M.S. *Challenger*, refitted with laboratories. They sailed nearly 70,000 nautical miles around the world, took soundings and water samples at hundreds of stops along the way, and discovered more than 4,000 new marine species. Noted for the discovery of the Mid-Atlantic Ridge and the Pacific's deepest trench, the expedition laid the foundations for modern oceanography. This acclaimed two-volume account, first published in 1877, summarises the major discoveries for the Atlantic legs of this pioneering voyage. Illustrated with plates and woodcuts, Volume 2 describes the voyage from the Caribbean via Madeira to the coast of Brazil, then to South Africa. The voyage home in 1876 from the Strait of Magellan is also covered. A final chapter summarises the principal conclusions.

Cambridge University Press has long been a pioneer in the reissuing of out-of-print titles from its own backlist, producing digital reprints of books that are still sought after by scholars and students but could not be reprinted economically using traditional technology. The Cambridge Library Collection extends this activity to a wider range of books which are still of importance to researchers and professionals, either for the source material they contain, or as landmarks in the history of their academic discipline.

Drawing from the world-renowned collections in the Cambridge University Library and other partner libraries, and guided by the advice of experts in each subject area, Cambridge University Press is using state-of-the-art scanning machines in its own Printing House to capture the content of each book selected for inclusion. The files are processed to give a consistently clear, crisp image, and the books finished to the high quality standard for which the Press is recognised around the world. The latest print-on-demand technology ensures that the books will remain available indefinitely, and that orders for single or multiple copies can quickly be supplied.

The Cambridge Library Collection brings back to life books of enduring scholarly value (including out-of-copyright works originally issued by other publishers) across a wide range of disciplines in the humanities and social sciences and in science and technology.

The Atlantic

*A Preliminary Account of the General Results
of the Exploring Voyage of H.M.S. Challenger
During the Year 1873
and the Early Part of the Year 1876*

VOLUME 2

CHARLES WYVILLE THOMSON

CAMBRIDGE
UNIVERSITY PRESS

CAMBRIDGE
UNIVERSITY PRESS

University Printing House, Cambridge, CB2 8BS, United Kingdom

Cambridge University Press is part of the University of Cambridge.

It furthers the University's mission by disseminating knowledge in the pursuit of
education, learning and research at the highest international levels of excellence.

www.cambridge.org
Information on this title: www.cambridge.org/9781108074759

© in this compilation Cambridge University Press 2014

This edition first published 1877
This digitally printed version 2014

ISBN 978-1-108-07475-9 Paperback

𝕿𝖍𝖊 𝖁𝖔𝖞𝖆𝖌𝖊 𝖔𝖋 𝖙𝖍𝖊 '𝕮𝖍𝖆𝖑𝖑𝖊𝖓𝖌𝖊𝖗,'

THE ATLANTIC.

CONTOUR MAP OF THE ATLANTIC

From Soundings and Temperature Observations up to May, 1876.

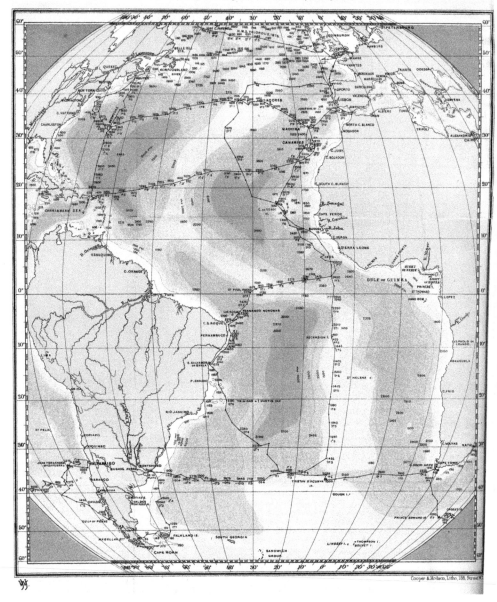

The Voyage of the 'Challenger.'

THE ATLANTIC

A PRELIMINARY ACCOUNT OF THE GENERAL RESULTS

OF

THE EXPLORING VOYAGE OF H.M.S. 'CHALLENGER'

DURING THE YEAR 1873

AND THE EARLY PART OF THE YEAR 1876

BY

SIR C. WYVILLE THOMSON

KNT., LL.D., D.Sc., F.R.SS.L. & E., F.L.S., F.G.S., Etc.

Regius Professor of Natural History in the University of Edinburgh,
And Director of the Civilian Scientific Staff of the 'Challenger' Exploring Expedition.

IN TWO VOLUMES.

VOLUME II.

Published by Authority of the Lords Commissioners of the Admiralty.

London:

MACMILLAN AND CO.

1877.

LONDON
R. CLAY, SONS, AND TAYLOR,
BREAD STREET HILL, E.C.

CONTENTS.

CHAPTER III.

BAHIA TO THE CAPE.

CHAPTER IV.

THE VOYAGE HOME.

CHAPTER V.

GENERAL CONCLUSIONS.

LIST OF ILLUSTRATIONS.

WOODCUTS.

VIGNETTES.

EXPLANATION OF THE PLATES.

THE ATLANTIC.

THE ATLANTIC.

CHAPTER I.

BERMUDAS TO MADEIRA.

Final departure from Bermudas. — Temperature-sounding near the Island.—*Scalpellum regium.*—Manganese concretions.—Gulf-weed. —*Gnathophausia.*—The general aspect of the Açores.—The Island of Fayal.—Horta.—The Island of Pico.—San Miguel.—Ponta Delgada.—Orange cultivation.—An excursion to Furnas.—Sete-Cidades.—The garden of M. José do Canto.—A religious ceremony.—*Flabellum alabastrum.*—*Ceratotrochus nobilis.*—Arrival at Funchal.

APPENDIX A.—Table of Temperatures observed between Bermudas and Madeira.
APPENDIX B.—-Table of Specific Gravities observed between Bermudas and Madeira.

WE left Bermudas on Thursday, the 12th of June, for the Açores. His Excellency Major-General Lefroy, C.B., F.R.S., Governor of the island, with his private secretary Captain Trench, and Captain Aplin, R.N. Captain-Superintendent of the Dockyard, and a party of ladies came on board in the afternoon; and we bade farewell with great regret to the friends from whom we had received such unvarying kindness

Temp.		Fms.
25°C		Surf
19°		75
18°		130
17°		500
16°		530
12°		370
10°		400
		440
8°		500
7°		530
6°		560
5°		600
4°		700
3°		1100
1·65°		2360

FIG. 1.—Diagram constructed from
Serial Sounding No. 59.

during our stay. At half-past
five we steamed out of the
Camber and passed among the
reefs to Murray's anchorage
on the north-east side of the
island, where we anchored for
the night. Next morning we
proceeded through the narrows,
and early in the forenoon, having seen the last of the beautiful though treacherous purple
shadows in the bright green
waters of Bermudas, we set all
plain sail and stood on our
course to Fayal. In the afternoon we got up steam and
sounded, lat. 32° 37′ N., long.
64° 21′ W., in 1,500 fathoms,
with the usual grey-white chalky
bottom which surrounds the
reefs. A serial temperature-
sounding indicated a distribution of temperature very similar to that at Station 55 to the
north of Bermudas; the warm
band to a depth of 350 fathoms
was still very marked (Fig. 1).

Our position at noon on the
15th was, lat. 33° 41′ N., long.
61° 28′ W., 1,610 miles from
Fayal.

On the morning of the 16th
we sounded in 2,575 fathoms
with a bottom of reddish ooze

containing many foraminifera. The bottom-temperature was 1°·5. A small, rather heavy trawl, with a
beam 11½ feet in length, was put over in the morning;
but when it was hauled in about five in the afternoon
it was found that it had not reached the bottom. This
was the first case of failure with the trawl; it was probably caused by the drift of the ship being somewhat
greater than we supposed. The net contained a specimen of one of the singular and beautiful fishes belonging to the Sternoptychidæ, an aberrant family of the
Physostomi, distinguished by having on some part
of the body ranges of spots or glands producing a
phosphorescent secretion. The surface of the body
is in most of the species devoid of scales, but in lieu
of these the surface of the skin is broken up into
hexagonal or rectangular areæ, separated from one
another by dark lines and covered with brilliant
silvery pigment dashed with various shades of bronze,
or green, or steel blue. We have taken in all five or
six species of these fishes in the net when dredging
or trawling. They certainly, however, do not come
from the bottom; it seems probable that they are
caught in the net on its passage at some little distance
below the surface, where there is reason to believe
that there is a considerable development of a peculiar
pelagic fauna.

On the 17th the trawl was lowered at seven in the
morning, and in the afternoon a sounding was taken
in 2,850 fathoms. Several examples of a large and
handsome species of the genus *Scalpellum* came up
in the trawl, a few still adhering to some singular
looking concretionary masses which they brought up
along with them.

Scalpellum regium (Fig. 2) is one of the largest of the known living species of the genus. The extreme length of a full sized specimen of the female is 60 mm., of which 40 mm. are occupied by the capitulum and 20 mm. by the peduncle. The capitulum is much compressed, 25 mm. in width from the occludent

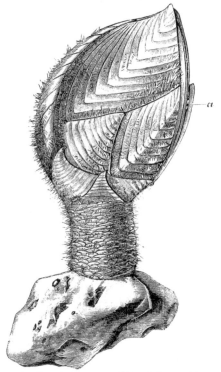

FIG. 2.—*Scalpellum regium* WYVILLE THOMSON. Natural size. *a.* Males lodged within the edge of the scutum. (No. 63.)

margin of the scutum to the back of the carina. The valves are 14 in number; they are thick and strong, with the lines of growth strongly marked, and they fit very closely to one another, in most cases slightly overlapping. When living, the capitulum is covered

with a pale brown epidermis, with scattered hairs of
the same colour.

The scuta are slightly convex, nearly once and a
half as long as broad. The upper angle is considerably prolonged upwards and, as in most fossil species,
the centre of calcification is at the upper apex. A
defined line runs downwards and backwards from the
apex to the angle between the lateral and basal
margins. The occludent margin is almost straight;
there is no depression for the adductor muscle, and
there is no trace of notches or grooves along the
occludent margin for the reception of the males; the
interior of the valve is quite smooth. The terga are
large, almost elliptical in shape, the centre of calcification at the upper angle. The carina is a handsome
plate, very uniformly arched, with the umbo placed
at the apex; two lateral ridges and a slight median
ridge run from the umbo to the basal margin; the
lower part of the valve widens out rapidly and
the whole is deeply concave. The rostrum, as in
Scalpellum vulgare, is very minute, entirely hidden
during life by the investing membrane. The upper
latera are triangular, the upper angle curving rather
gracefully forwards; the umbo of growth is apical.

The rostral latera are long transverse plates lying
beneath the basal margins of the scuta. The carinal
latera are large and triangular, with the apex curved
forwards very much like the upper latera, and the
infra-median latera are very small, but in form and
direction of growth nearly the same.

The peduncle is round in section and strong, and
covered with a felting of light-brown hair. The scales
of the peduncle are imbricated and remarkably

large, somewhat as in *S. ornatum* DARWIN. About
three, or at most four scales, pass entirely round
the peduncle. The base of attachment is very small,
the lower part of the peduncle contracting rapidly.
Some of the specimens taken were attached to
the lumps of clay and manganese concretions, but
rather feebly, and several of them were free and
showed no appearance of having been attached.
There is no doubt, however, that they had all been
more or less securely fixed, and had been pulled from
their places of attachment by the trawl. On one
lump of clay there were one mature specimen and
two or three young ones, some of these only lately
attached. The detailed anatomy of this species will
be given hereafter, but the structure of the soft parts
is much the same as in *Scalpellum vulgare.*

In two specimens dissected there was no trace of
a testis or of an intromittent organ, while the ovaries
were well developed ; I conclude, therefore, that the
large attached examples are females, corresponding,
in this respect, with the species otherwise also most
nearly allied, *S. ornatum.*

In almost all the specimens which were procured
by us, several males, in number varying from five to
nine, were attached within the occludent margins of
the scuta, not embedded in the chitinous border of
the valve, or even in any way in contact with the
shell, but in a fold of the body-sac quite free from
the valve. They were ranged in rows, sometimes
stretching—as in one case where there were seven
males on one side—along the whole of the middle
two-thirds of the edge of the tergum.

The male of *Scalpellum regium* (Fig. 3) is the

simplest in structure of these parasitic males which
has yet been observed. It is oval and sac-like, about
2 mm. in length by ·9 mm. in extreme width. There
is an opening at the upper extremity which usually
appears narrow, like a slit, and this is surrounded
by a dark, well-defined, slightly raised ring. The
antennæ are placed near the posterior extremity of
the sac, and resemble closely in form those of
S. vulgare. The whole of the sac, with the exception
of a small bald patch near the point of attachment,
is covered with fine chitinous hairs
arranged in transverse rings. There
is not the slightest rudiment of a
valve, and I could detect no trace of
a jointed thorax, although several
specimens were rendered very transparent by boiling in caustic potash.
There seems to be no œsophagus nor
stomach, and the whole of the posterior two-thirds of the body in the
mature specimens was filled with a
lobulated mass of sperm-cells. Under
the border of the mantle of one female
there were the dead and withered remains of five males, and in most cases
one or two of the males were not fully developed ;
several appeared to be mature, and one or two were
dead,—empty, dark-coloured chitine sacs.

Fɪɢ 3.—Male of *Scalpel-
lum regium.* Twenty
times the natural size.
(No. 63.)

The concretionary masses to which the barnacles
adhered were irregular in form and size. One, for
example, to which a large *Scalpellum* was attached,
was irregularly oval in shape, about three centimetres
in length and two in width. The surface was mam-

millated and finely granulated, and of a dark brown colour, almost black. A fracture showed a semi-crystalline structure; the same dark-brown material arranged in an obscurely radiating manner from the centre and mixed with a small quantity of greyish-white clayey matter. This nodule was examined by Mr. Buchanan, and found to contain, like the nodules dredged in 2,435 fathoms at Station 16, 700 miles to the east of Sombrero, a large percentage of peroxide of manganese. Some other concretionary lumps were of a grey colour, but all of them contained a certain proportion of manganese, and they seemed to be gradually changing into nodules of pyrolusite or wad by some process of infiltration or substitution.

On Wednesday, June 18, we resumed our course with a fine breeze, force 5 to 7, from the south-east. In this part of our voyage we were again greatly struck with the absence of the higher forms of animal life. Not a sea-bird was to be seen, with the exception of a little flock of Mother Carey's chickens, here apparently always *Thalassidroma wilsoni*, which kept playing round the ship on the watch for food, every now and then concentrating upon some peculiarly rich store of offal as it passed astern, and staying by it while the ship went on for a quarter of a mile, fluttering above the water and daintily touching it with their feet as they stooped and picked up the floating crumbs, and then rising and scattering in the air to overtake us and resume their watch.

The sea itself in the bright weather, usually under a light breeze, was singularly beautiful—of a splendid indigo-blue of varying shades as it passed from sunlight into shadow, flecked with curling white crests;

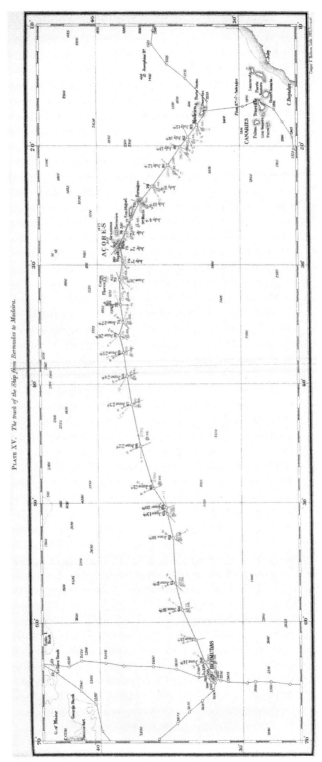

PLATE XV. *The track of the Ship from Bermudas to Madeira.*

The material originally positioned here is too large for reproduction in this reissue. A PDF can be downloaded from the web address given on page iv of this book, by clicking on 'Resources Available'.

but it was very solitary : day after day went by
without a single creature—shark, porpoise, dolphin,
or turtle—being visible. Some gulf-weed passed from
time to time, and bunches of a species of *Fucus*, either
F. nodosus or a very nearly allied form, evidently
living and growing, and participating in the wandering and pelagic habits of *Sargassum*. The floating
islands of the gulf-weed, with which we had become
very familiar as we had now nearly made the circuit
of the ' Sargasso Sea,' are usually from a couple of
feet to two or three yards in diameter, sometimes
much larger ; we have seen on one or two occasions
fields several acres in extent, and such expanses are
probably more frequent nearer the centre of its area
of distribution.

They consist of a single layer of feathery bunches
of the weed (*Sargassum bacciferum*), not matted but
floating nearly free of one another, only sufficiently entangled for the mass to keep together.
Each tuft has a central brown thread-like branching
stem studded with round air-vesicles on short stalks,
most of those near the centre dead, and coated with
a beautiful netted white polyzoon. After a time
vesicles so encrusted break off, and where there is
much gulf-weed the sea is studded with these little
separate white balls. A short way from the centre,
towards the ends of the branches, the serrated willow-
like leaves of the plant begin; at first brown and
rigid, but becoming farther on in the branch paler,
more delicate, and more active in their vitality. The
young fresh leaves and air-vesicles are usually ornamented with the stalked vases of a *Campanularia*.
The general colour of the mass of weed is thus olive

in all its shades, but the golden olive of the young and growing branches greatly predominates. This colour is, however, greatly broken up by the delicate branching of the weed, blotched with the vivid white of the encrusting polyzoon, and riddled by reflections from the bright blue water gleaming through the spaces in the network. The general effect of a number of such fields and patches of weed, in abrupt and yet most harmonious contrast with the lanes of intense indigo which separate them, is very pleasing.

These floating islands have inhabitants peculiar to them, and I know of no more perfect example of protective resemblance than that whch is shown in the gulf-weed fauna. Animals drifting about on the surface of the sea with such scanty cover as the single broken layer of the sea-weed, must be exposed to exceptional danger from the sharp-eyed sea-birds hovering above them, and from the hungry fishes searching for prey beneath; but one and all of these creatures imitate in such an extraordinary way, both in form and colouring, their floating habitat, and consequently one another, that we can well imagine their deceiving both the birds and the fishes. Among the most curious of the gulf-weed animals is the grotesque little fish *Antennarius marmoratus* (Fig. 44, vol i., p. 95), which finds its nearest English ally in the 'fishing frog' (*Lophius piscatorius*), often thrown up on the coast of Britain, and conspicuous for the disproportionate size of its head and jaws, and for its general ugliness and rapacity. None of the examples of the gulf-weed *Antennarius* which we have found are more than 50 mm. in length, and we are still uncertain whether such individuals have attained their

full size. It is this little fish which constructs the
singular nests of gulf-weed bound in a bundle with
cords of a viscid secretion which have been already
mentioned as abundant in the path of the Gulf-stream.

Scillæa pelagica, one of the shell-less mollusca, is
also a frequent inhabitant of the gulf-weed. A little
short-tailed crab (*Nautilograpsus minutus*) swarms
on the weed and on every floating object, and it is
odd to see how the little creature usually corresponds
in colour with whatever it may happen to inhabit.
These gulf-weed animals, fishes, mollusca, and crabs,
do not simply imitate the colours of the gulf-weed;
to do so would be to produce suspicious patches of
continuous olive; they are all blotched over with
bright opaque white, the blotches generally rounded,
sometimes irregular, but at a little distance absolutely
undistinguishable from the patches of *Membranipora*
on the weed. Mr. Murray, who has the general
superintendence of our surface work, brings in
curious stories of the habits of the little crabs. He
observes that although every floating thing upon the
surface is covered with them, they are rarely met with
swimming free, and that whenever they are dislodged
and removed a little way from their resting-place
they immediately make the most vigorous efforts to
regain it. The other day he amused himself teasing
a crab which had established itself on the crest of a
Physalia. Again and again he picked it off and put
it on the surface at some distance, but it always
turned at once to the *Physalia* and struck out, and
never rested until it had clambered up into its former
quarters.

On Thursday, the 19th, we sounded in 2,750

fathoms in a grey mud containing many foraminifera. Position of the ship at noon, Lat. 35° 29′ N., Long. 50° 53′ W.

The wind now gradually freshened, and for the next three days we went on our course with a fine breeze, force from 4 to 7, from the southward, sounding daily at a depth of about 2,700 fathoms, with a bottom of reddish-grey ooze. On Tuesday the 24th the trawl was put over in 2,175 fathoms, Lat. 38° 3′ N., Long. 39° 19′ W., about 500 miles from the Açores. As in most of the deep trawls on grey mud, a number of the zoœcia of delicate branching polyzoa were entangled in the net. One of these on this occasion was very remarkable from the extreme length (4 to 5 mm.) of the pedicels on which its avicularia were placed. Another very elegant species was distinguished by the peculiar sculpture of the cells, reminding one of those of some of the more ornamented *Lepraliæ*.

On Wednesday the 25th a serial sounding (Fig. 4) showed that the layer of warm water which envelops Bermudas was gradually thinning out and disappearing, and a sounding on the 27th (Fig. 5) brought out the same result even more clearly, the isotherm of 16° C. which at Station 59 was at a depth of 330 fathoms having now risen to 50 fathoms below the surface.

On Monday the 30th of June we sounded in 1,000 fathoms, about 114 miles westward from Fayal. The dredge was put over early in the forenoon, and came up half filled with a grey ooze with a large proportion of the dead shells of pteropods, many foraminifera, and pebbles of pumice. Many animal forms of great

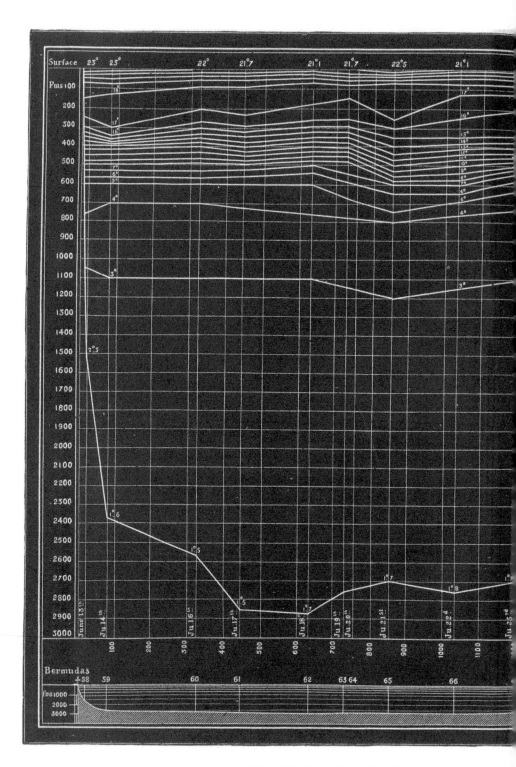

PLATE XVI.—DIAGRAM OF THE VERTICAL DISTRIBUTION

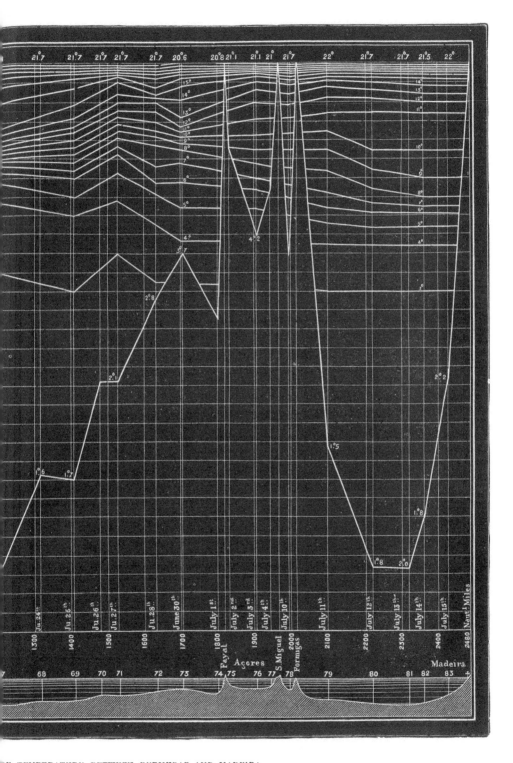

F TEMPERATURE BETWEEN BERMUDAS AND MADEIRA.

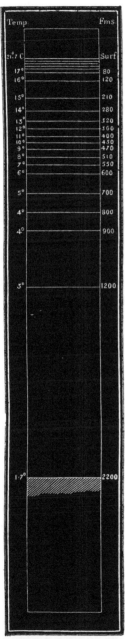

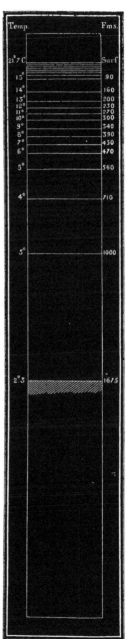

FIG. 4.—Diagram showing the relation
between depth and temperature at
Station 69.

FIG. 5.—Diagram showing the relation
between depth and temperature at
Station 71.

interest were found entangled in the swabs, or sifted out of the mud. A schizopod crustacean of large size, and great beauty of form and brilliancy of colouring, came up in this haul. Dr. von Willemoes-Suhm regards it as congeneric with the species taken at Station 69 at a depth of 2,200 fathoms, and as these crustaceans are among our most interesting acquisitions during the voyage between Bermudas and the Açores, I will abstract a brief description of them from his notes.

The two crustaceans for whose reception Dr. von Willemoes-Suhm proposes to establish the genus *Gnathophausia* present characters which have hitherto been found partly in schizopods and partly in phyllopods, but not combined in the same animal. They are, however, essentially schizopods, and have much in common with *Lophogaster*, a genus described in great detail by the late Prof. Sars. It is proposed to refer *Gnathophausia* to the family LOPHOGASTRIDÆ, which must be somewhat modified and expanded for its reception.

In *Gnathophausia* the dorsal shield covers the thoracic segments of the body, but it is unconnected with the last five of these. The shield is prolonged anteriorly into a spiny rostrum. The stalked eyes are fairly developed in the ordinary position. There is an auxiliary eye on each of the maxillæ of the second pair.

The two species of the genus are thus distinguished : *G. gigas*, n. sp. (Fig. 6). Scale of the outer antenna with five teeth ; dorsal shield with the outer angles of its posterior border produced into spines ; no posterior spine in the middle line ; length 142 mm.

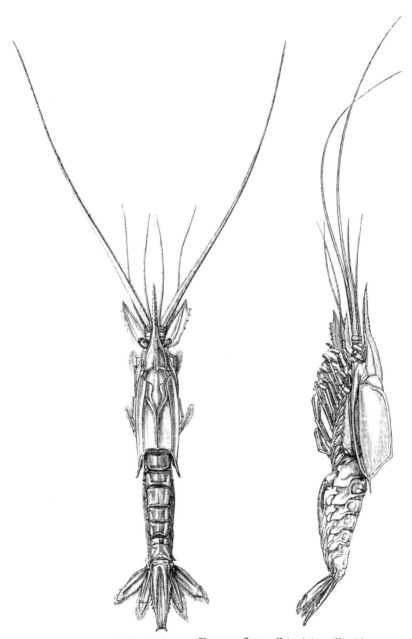

Fig. .—*Gnathophausia gigas*, von WILLEMOES-SUHM. Natural size. (No. 69.)

Of this species one specimen was procured from a
depth of 2,200 fathoms with a bottom of globigerina-
ooze at Station 69, 400 miles to the west of the
Açores.

Gnathophausia Zoëa (Fig. 7.) has the scale of the

FIG. 7.—*Gnathophausia Zoëa*, von WILLEMOES-SUHM. Natural size. (No. 73.)

outer antenna with one tooth ; a long central spine
on the posterior border of the dorsal shield, but no
lateral spines ; length 60 mm.   A single specimen of
this species occurred (Station 73).   On comparing the
figures of these two species and of their anatomical
details with those of *Lophogaster* given by Sars, one
is struck by their great general similarity ; but there
are characters presented by the new genus, particularly in connection with the dorsal shield, which
not only entirely separate it from *Lophogaster* but
enlarge our views on the whole schizopod group.   In
both species the shield is sculptured by ridges
traversing it in different directions, and in both there
is a long spiny rostrum ; but this shield is merely a
soft duplicature of the skin connected with the body
only anteriorly, and leaving five thoracic segments
entirely free.   In the structure of the shield and
in its mode of attachment *Gnathophausia* has the
greatest resemblance to *Apus* among all crustaceans,
but it differs from it widely in other respects.
Nebalia is the only schizopod in which the carapace

is not connected with the posterior thoracic segments, but in that genus the form of the carapace is totally different, and the genera are otherwise in no way nearly related. Neither the antennæ, nor the scales, nor the parts of the mouth, present any marked differences from those of *Lophogaster*, with the exception of the second maxillæ. These, with nearly the same form as in the Norwegian genus, bear a pair of accessory eyes. Such eyes are well known at the base of the thoracic and even of the abdominal limbs in the EUPHAUSIDÆ, a family with which the LOPHOGASTRIDÆ have otherwise nothing in common, but hitherto they have not been met with in any other animal on any of the manducatory organs.

Of the eight pairs of legs seven are ambulatory, only the first pair is, as in *Lophogaster*, transformed into maxillipeds. The gills are arborescent and attached to the bases of the legs. The abdomen and its appendages scarcely differ from those of *Lophogaster*. We find here also that the last segment is apparently divided into two. This would indicate an approach to such forms as *Nebalia*, which has nine abdominal segments, or at all events a tendency to a multiplication of segments which if really existing would scarcely allow the association of the genus with the true schizopods.

The weather was remarkably fine. During the day the Island of Flores was visible like a cloud on the horizon about fifty miles to the northward. As our sounding was comparatively shallow our position was probably on a southern extension of the rise which culminates northwards in Flores and Corvo. One of the most remarkable differences between the Açores and Bermudas is that, while Bermudas springs up, an isolated peak, from a great depth, the Açores seem to be simply the highest points of a great plateau-like elevation, which extends for upwards of a thousand miles from west to east, and appears to be continuous with a belt of shallow water stretching to Iceland in the north, and connected probably with the 'Dolphin Rise' to the southward—a plateau which in fact divides the North Atlantic longitudinally into two great valleys, an eastern and a western. The three previous soundings, the first three hundred and thirty miles from Fayal, had already shown that we were passing over the gradual ascent; and this dredging, although not very fruitful in results, gave indications, by the presence of some comparatively shallow-water northern species, of a northern extension of its conditions.

Although the two remote little archipelagos out in the Atlantic have many things in common, the first impression of the Açores is singularly different from that of Bermudas. Long before the white cottages, straggling in broken lines almost round the islands on the top of the sea-cliff, or grouped in villages round their little churches—white, quaintly edged with black, like mourning envelopes—in the mouths of richly-wooded ravines, have become visible; the eye has been dwelling with pleasure on the bold outline of the land, running up everywhere into magnificent ridges and pinnacles, and has sometimes been almost startled by the sudden unveiling of a majestic peak through a rift in the clouds far up in the sky.

As the islands draw nearer, the hazy blues and purples give place to vivid shades of green, and these,

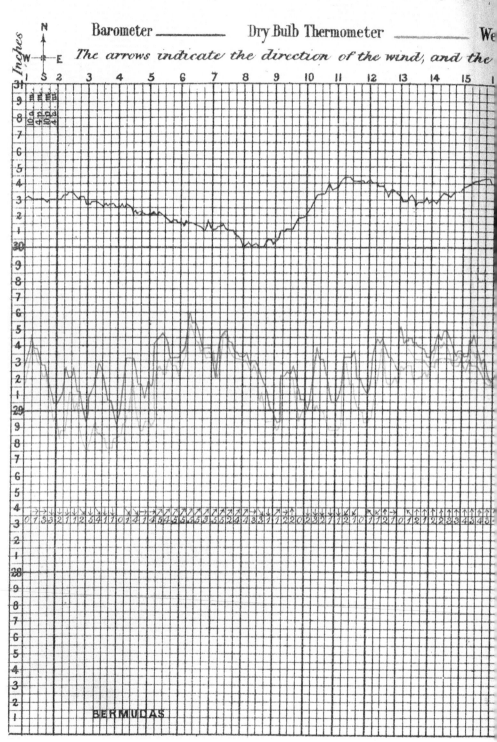

PLATE XVII. *Meteorological Ob*

Barometer ———— Dry Bulb Thermometer ———— We

The arrows indicate the direction of the wind, and the

BERMUDAS

Bulb Thermometer ⎯⎯⎯ Temperature of Sea Surface ⎯⎯⎯

umbers beneath its force according to Beaufort's scale

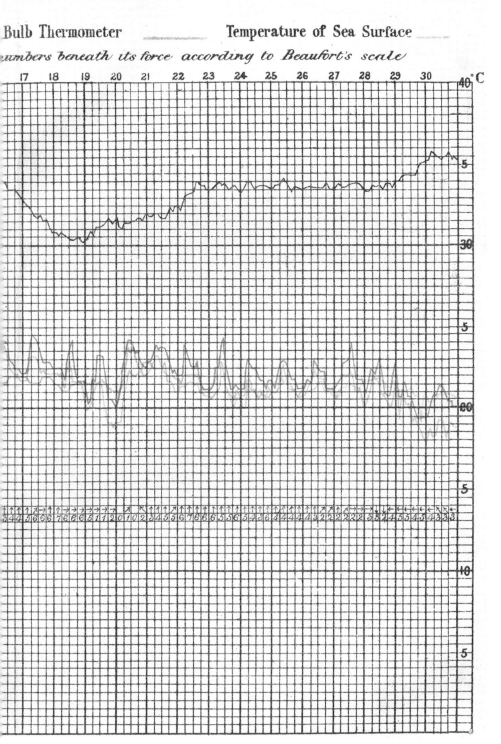

at first blending and indistinct, gradually resolve themselves into a rich and luxuriant cultivation, filling up the valleys with dense woods of pine and sweet chestnut, covering the slopes wherever it is possible to work, with orange-groves, vineyards, and fields of maize and wheat; and even where work is scarcely possible, mitigating the nearly precipitous cliffs and rendering them available by means of artificial terraces and slopes.

The Açores are situated between lat. 39° 45′ and 36° 50′ N., and long. 25° and 31° 20′ W. They consist of three groups,—two small islands, Flores and Corvo, to the extreme north-west; Fayal and Pico, separated by a narrow and shallow channel and forming geologically one elevation, in the centre; and associated with these, spreading to the north-eastward, San Jorge, Terceira the former seat of government, and Graciosa; while the third group, nearly a couple of hundred miles to the south-westward, consists of San Miguel the richest and most important of the islands and the seat of the present capital, San Maria, and two curious little patches of naked rocks the Formigas and Dollabarat. The climate of the Açores is mild and equable. In summer they are touched by the south-east trades, or rather they are just in the fine-weather edge of the variables. In winter they are subject to heavy gales from the south-west. Their climate is doubtless influenced to a certain extent by the southern deflection of the Gulf Stream, and they are near the northern border of the Sargasso Sea. The mean annual temperature of Ponta Delgada in San Miguel is 17°·67 C., 0°·9 higher than that of Palermo, 1°·4 lower than that of Malaga, and 0°·6 lower than that of Funchal. The mean winter temperature of Ponta Delgada is 13°·05 C., 1°·8 higher than that of Palermo, and 2°·7 lower than that of Funchal; and the mean summer temperature is 20°·67, 1°·3 lower than that of Palermo, and 0°·2 lower than that of Funchal. The mean temperature of the warmest month at San Miguel is 22°·67 C., and that of the coldest 12°·28 ; the range between the extremes is therefore only about 10° C.

All the islands are volcanic, and their structure recalls, in every respect, that of such comparatively modern volcanic districts as those of the Eifel or Auvergne. The high rugged crests, which everywhere take the form of more or less complete amphitheatres, are the walls of ancient craters, the centres of earlier volcanic action. The bottom of the old crater is now usually occupied by a lake, and in it, or round its edges, or outside it on its flanks, there are often minor craters, frequently very perfect in their form, which indicate eruptions of later date, efforts of the subsiding fires. The rocks which everywhere stretch down in great undulating masses from the sides of the craters to the sea, are lavas of different dates, some of them not much more than a century old; the wooded ravines are sometimes the natural intervals between lava streams, deepened by rivulets which have naturally followed their direction; more frequently they are valleys of erosion, worn by torrents in intervening accumulations of loose scoriæ; and the splendid cliffs, which form an inaccessible wall round the greater part of most of the islands, show, in most instructive sections, the basaltic,

trachytic, and trachydoleritic lavas, and the rudely
or symmetrically stratified sub-aërial or sub-marine
beds of tufa and ashes, the products of successive
eruptions. As a rule, soil formed by the wearing
down of volcanic rocks is highly favourable to the
growth of plants. It is wonderful to see how the
coulées of lava and the mounds of pumice and ashes,
formed by even the most recent eruptions—of many
of which we know the dates, such as those of 1512,
1672, 1718, and 1722—are now covered with corn-
fields and vineyards, and, in inaccessible places, with
a luxuriant native vegetation.

Next day we sounded in thirteen hundred and fifty
fathoms, about twenty miles west of Fayal, in the
depression which separates the western from the
central group, and during the afternoon the fine bold
island approached us, alternate cloud and bright
sunshine bringing out the full effects of its contour
and colouring. The south coast of Fayal is bounded
by an abrupt cliff, perhaps from one to two hundred
feet in height, intersected every here and there by
deep valleys, and showing, where the cliff is too
precipitous to support vegetation, sections of lava
streams of various colours, and of beds of irregularly
stratified scoriæ and ashes. The main road runs
along the top of the cliff, and at intervals, usually at
the point of intersection of a wooded gorge, a village
of low white cottages clusters round a black-and-
white church, surmounted by a large black cross.
From the road thê land slopes gradually upwards,
passing into wide valleys terminating in ravines in
the side of the Caldeira, a peak upwards of three
thousand feet in height, with a fine crater at the top

of it, not far from the centre of the island; or
running up abruptly upon the many secondary cones
and mounds of scoriæ which are scattered in all
directions. On this side of the island wheat is chiefly
cultivated, except in the valleys, where there are a
few vineyards and fields of maize upon the slopes.
The wheat was already yellowing for the harvest.
The fields are small, separated by walls of dark
lava, built, as we were afterwards told, partly as
shelter from the high winds, and partly as the easiest
means of stowing the lava blocks, which have to be
removed from the ground in the process of clearing.
Another very effective addition to the fence serves
also a double purpose, a hedge of the common reed
(*Arundo donax*) is usually planted within the wall,
and runs up to a height of twelve or fifteen feet,
adding greatly to the shelter, and producing a long,
straight, light cane, which is used in many ways;
split up, it answers the purpose of laths for supporting
plaster, and the round canes, bound together and
often fitted in neat patterns, may often be seen in
the peasants' houses forming partitions, cupboards,
or light odds and ends of furniture. These tall
reed hedges, at this season bearing large, feathery
flower-heads on this year's shoots, while the stems
of last year, now becoming hard and woody, bear on
side branches a crop of small leaves like those of
the bamboo, form quite a peculiar feature in the
landscape. The Caldeira itself, the father of the
family of craters, and evidently the centre of the
first and most powerful outburst of volcanic action,
remained invisible to us—shrouded all day under a
thick canopy of cloud.

In the evening we steamed into the channel between Fayal and Pico, and anchored in the road-stead of Horta, the chief town of Fayal. Here we were visited by the Portuguese officer of health, who, while making strict inquiries as to the presence of contagious disease in the ports which we had previously visited, said nothing about the health of his own town; and it was with extreme chagrin that we learned from the British Vice-Consul, who came on board shortly afterwards, that Horta was suffering from an epidemic of small-pox, which had latterly been rather severe, especially among children. Under these circumstances Captain Nares judged it imprudent to give general leave, and on that evening and on the following morning one or two of us only took a rapid run through the town and its immediate neighbourhood, to gain such a hasty impression as we might of its general effect.

Horta is a pretty little town of ten thousand inhabitants, situated in a deep bay which opens to the westward, and looks straight across to the island of Pico, distant about four miles. The bay is bounded to the north by a bold lava promontory, Ponta Espalamaca; and on the south by a very remarkable isolated crater, with one half of its bounding wall broken down and allowing the sea to enter, called Monte da Guia, a very prominent object when entering the bay from the southward. Monte da Guia is almost an island, and apparently at one time it was entirely detached. It is now connected with the land by a narrow neck, composed chiefly of soft scoriæ and pumice, in the middle of which there juts up an abrupt mass of dark rock called ‘Monte

Queimada' (the burnt mountain), formed partly of
stratified tufa of a dark chocolate colour, and partly
of lumps of black lava, porous, and each with a large
cavity in the centre, which must have been ejected as
volcanic bombs in a glorious display of fireworks at
some period beyond the records of Açorean history,
but late in the geological annals of the islands.

A long straggling street follows the curve of the
bay and forks into two at the northern end, and
cross streets ending in roads bounded by high sheltering walls, many of them white, tastefully relieved
with blue or grey simple frieze-like borders, run up
the slope into the country. The streets are narrow,
with heavy green verandahs to the houses, and have
a close feel; but the town is otherwise clean and
tidy, the houses are good in the ordinary Portuguese
style, and some of the convent churches, though
ordinary in their architecture, are large and even
somewhat imposing. The church of the monastery
occupied by the Carmelites before the suppression of
the religious orders, overlooking the town, with its
handsome façade surmounted by three Moresque
cupolas, is the most conspicuous of these; and the
Jesuit church, built somewhat in the same style, a
little farther back from the town, is also rather
effective. The suburbs abound in beautiful gardens;
but they are surrounded by envious walls, and the
unfortunate circumstances of our visit prevented our
making the acquaintance of their possessors, of
whose friendly hospitality we had heard much.

Pico, facing the town at the opposite side of the
narrow strait, is at once a shelter to Horta and a
glorious ornament. The peak, a volcanic cone of

7,613 feet in height,- rivals Etna or the Peak of Teneriffe in symmetry of form. The principal cone terminates in a crater about two hundred feet deep, and nearly in the centre of the great crater a secondary cone, very perfect in shape, and composed of scoriæ and lava, rises to a height of upwards of two hundred feet above its rim. This little additional peak gives the top of this mountain a very characteristic form. The top of the mountain is covered with snow during the winter months, but it has usually entirely disappeared before the end of May. The sides of the mountain, alternately ridged and deeply grooved, and studded with the cones and craters of minor vents, are richly wooded, and the lower and more level belt sloping down to the sea-cliff, produces abundance of maize, yams (*Calocasia esculenta*), and wheat. The other islands depend greatly upon Pico for their supply of vegetables, fruit, and poultry. The morning we were at Fayal a fleet of Pico boats, two-masted with large latteen sails, loaded with green figs, apricots, cabbages, potatoes, and fowls, crossed over in time for early market. Formerly Pico was the vineyard of the Açores. Previous to the year 1853 from twenty to thirty thousand pipes were exported from the island of a dry, rather high-flavoured wine, which commanded a fair price in the markets of Europe, under the name of 'Pico Madeira.' In 1853 the wretched *Oidium tuckeri* devastated the vineyards and reduced the population of the island, who depended mainly on their wine production for their subsistence, to extreme misery. Nothing would stop the ravages of the fungus; in successive years the crop was

reduced to one-fourth, one-eighth, one-tenth, and then
entirely ceased, and the inhabitants emigrated in
great numbers to Brazil and California. Some few
attempts have been made to restore the vines, but up
to the present time there is practically no manufac-
ture of wine in the Açores.

We left Fayal the morning after our arrival, and
had one or two hauls of the dredge in shallow water,
from fifty to a hundred fathoms, in the channel
between Fayal and Pico. Everywhere the bottom
gave evidence of recent volcanic action. The dredge
came up full of fine dark volcanic sand and pieces
of pumice. We were surprised to find the fauna varied
and abundant. As in the case of plants, it seems
in some cases to take but little time for animals to
spread in undiminished numbers over an area where
every trace of life must assuredly have been destroyed
by the rain of fire and brimstone. In the evening
we passed eastward through the channel between
Pico and San Jorge, and greatly enjoyed the fine
scenery of the latter island, which rises inland into
a bold mountain ridge, and presents to the sea a
nearly unbroken mural cliff, ranging to upwards of
five hundred feet in height.

On the evening of the 4th of July we anchored
in the roadstead of Ponta Delgada, the capital of
San Miguel, and the chief town of the Açores. We
were a little anxious about Ponta Delgada, for we
had been told at Fayal that small-pox was preva-
lent there also ; and although our information was
not very definite, and we were in hope that it
might prove incorrect, it was with great satisfaction that we heard from the quarantine officer

that they had had no cases for a year past. Leave was accordingly freely given, and we all prepared to make the most of our stay, which could not be extended beyond five days at the farthest if we hoped to hold to our future dates.

FIG. 8.—*Altingia excelsa,* in the garden of M. José do Canto, San Miguel.
(From a photograph.)

Ponta Delgada is very like Horta. It curves in the same way round the shore of a bay, and gardens and orange-groves clothe the slope of a receding amphitheatre of hills; but there is more space about it, and apparently more activity and enterprise. One of the first things we saw was a locomotive steam-engine bringing down blocks of lava, to satiate, if possible, the voracity of the sea, and enable them to finish in peace a very fine breakwater, for whose construction every box of oranges exported has paid a tax for some years past. The wild south-westerly storms of winter pull down the pier nearly as fast as it is built, and the engineer has adopted the plan of simply bringing an unlimited supply of rough blocks, and leaving the waves to work their wicked will with them and arrange them as they choose. In this way the blocks seem to be driven into the positions in which they can best resist the particular forces to which they are exposed, and they are subsiding into a solid foundation on which the building work is making satisfactory progress. Ponta Delgada is much larger than Horta; the streets are wider, and there are many more good-sized houses. The churches are numerous and large, but commonplace and immemorial; the only one which has any claim to a monumental character is an old church near the centre of the town, which was formerly attached to a Jesuit convent.

The market at Ponta Delgada does not appear to be very good, and, particularly in the short supply of vegetables and fruit, it seems to suffer from its distance from Pico.

On the morning of Saturday, the 5th of July, a merry party of about a dozen of us started from Ponta Delgada to see the celebrated valley and lake of the Furnas.

As the crow flies, the Furnas village, the fashionable watering-place of San Miguel, where the hot

springs and baths are, is not more than eighteen
miles from Ponta Delgada, but the road is circuitous
and hilly, and the entire distance to be gone over was
not much less than thirty miles. We engaged four
carriages, each drawn by three mules abreast, and
warranted to take us the whole distance, if we chose,
without drawing bridle.

Fig. 9.—*Cryptomeria japonica*, in the garden of M. Jose do Canto, San Miguel.
(From a photograph.)

The first part of our route lay through the long
drawn-out suburbs of the town, past one or two
churches without much character, very like those in
second-class towns in Spain and Portugal. We then
turned towards the interior, and walked up a long

ascent, not to harass our *mulos* so early in the journey.

The road was dreary and tantalizing. We knew that it was bordered by lovely orange-groves, the last of the fragrant flowers just passing over, and the young fruit beginning to swell, and usually about the size of a hazel-nut; but of this we saw nothing; our laborious climb was between two hot black walls of rough blocks of lava, nine to ten feet high. As a partial relief, however, a tall hedge of evergreen trees planted close within the walls rose high above them, and threw enough of shade to checker the glare on the dusty road beneath.

In the Açores at one time the orange-trees, which seem to have been introduced shortly after the discovery of the islands, were planted at a distance from one another, and allowed to attain their full size and natural form. Under this system some of the varieties formed noble trees with trunks eighteen inches in diameter. The wind-storms are, however, frequently very violent in winter, and often when the fruit was nearly ripe, the best part of a crop was lost, and the trees themselves greatly injured and broken by a south-westerly gale. Experience has now shown that larger crops may be procured with much greater certainty by dwarfing and sheltering the trees, and it has now become a nearly universal practice to surround the rectangular orchards or gardens, there called 'quintas,' with a lava wall; and further, to break the wind still more effectually, to plant within the wall a hedge of quickly-growing evergreen trees, which is allowed to overtop it by twenty feet or so, and to scatter tall evergreens wherever there is a clear space among the orange-trees, which are pruned and regulated so as to keep well below their level.

These tall hedges, intersecting the country in all directions, have a peculiar but rather agreeable effect. Almost all the hedge-plants are of a bright lively green. The one most used is *Myrica faya,* a native plant, which grows very abundantly on all the uplands, and seems to be regarded as a kind of badge in the islands, as its relation *Myrica gale* is in the west highlands of Scotland. Two other native plants, *Laurus canariensis* and *Persea indica,* are sometimes employed, but they are supposed to affect the soil prejudicially. Of late years a very elegant Japanese shrub, *Pittosporum undulatum,* which was originally introduced from England, has become widely used as a shelter-plant, and an allied species, *Pittosporum tobira,* is found to thrive well in quintas exposed to the sea-breeze.

It is needless to say that the culture of the orange is the main industry of San Miguel, and that the wonderful perfection at which this delicious fruit arrives has been sufficient to give the island an advantage over places less remote, and to ensure a reasonable amount of wealth to the owners of the ground. The cultivation of the orange is simple and inexpensive. The soil formed by the wearing down of the volcanic rocks is, as a rule, originally rich. It is inclosed and worked for a year or two, and young plants of good varieties, from layers or grafts, are planted at distances of eight or ten yards. Strong plants from layers begin to fruit in two or three years. They come into full bearing in from eight to ten years, when each tree should produce about

FIG. 10.—*Araucaria cookei*, in the garden of M. José do Canto, San Miguel.
(From a photograph.)

fifteen hundred oranges. The orange-trees are
lightly pruned, little more than the harsh spiny

shoots being removed. The surface of the ground is
kept clean and tidy with a hoe, and it is manured
yearly, or at longer intervals, by a method introduced
in old times into Britain by the Romans,—lupins,
which send up a rapid and luxuriant growth, and
produce a large quantity of highly nitrogenous seed
in the rich new soil, are sown thickly among the
trees, and then the whole, straw, pods, and seeds, are
dug into the ground. This seems to be sufficient to
mellow the soil, and any other manure is rarely used
for this crop.

The oranges begin to ripen early in November, and
from that time to the beginning of May a constant
succession of sailing vessels, and latterly steamers,
hurry them to the London market. The fruit is
gathered with great care, the whole population, old
and young, assisting at the harvest, and bringing it
down in large baskets to the warehouses in the town.
Each orange is then wrapped separately in a dry
maize leaf, and they are packed in oblong wooden
boxes, four to five hundred in the box. They used to
be packed in the large clumsy cases with the bulging
tops, so familiar in shops in England in the orange
season ; but the orange case has been entirely super-
seded during the last few years by the smaller box.
About half a million such boxes are exported yearly
from San Miguel, almost all to London. The prices
vary greatly. Oranges of the best quality bring
upon the tree 8*s.* to 15*s.* a thousand according to the
state of the market; and the expenses of gathering,
packing, harbour dues, and freight may come to 1*l.*
a thousand more ; so that counting the loss which
with so perishable a commodity cannot fail to be

considerable, each St. Michael's orange of good quality delivered in London costs rather more than a halfpenny. The price increases enormously as the season goes on. Several varieties are cultivated, and one variety ripens a comparatively small number of large fruit, without seeds, towards the middle of April, which bring sometimes ten times as much as the finest of the ordinary oranges in the height of the season.

At length, at an elevation of six hundred feet or so, the walls of the quintas were passed, and we emerged into the open country. The island is divided into two somewhat unequal portions, an eastern and a western. To the east we have high volcanic ridges, surrounding the picturesque valley of the Furnas, and stretching, in rugged peaks and precipitous clefts, to the extremity of the island. The western portion culminates in the Caldeira (or crater) of the Sete-Cidades, probably one of the most striking pieces of volcanic scenery to be met with anywhere.

Between the two there is a kind of neck of lower land, beds of lava and scoriæ and a congregation of small volcanic cones, wonderfully sharp and perfect, and with all the appearance of being comparatively recent. It is across this neck that the road passes to Furnas, and as it wound among the wooded dells between the cones we had a splendid view of the northern coast with its long line of headlands—lava flows separated by deep bays radiantly blue and white under the sun and wind and passing up into deep wooded dells. Beneath us, at the point where the road turned along the northern shore, lay the pretty little town of Ribeira Grande, the second on the island.

This middle belt of lower land is, perhaps, with

FIG. 11.—Orange Groves near Ponta Delgada. (From a Photograph.)

the exception of the land immediately round the towns, the best cultivated part of the island. The volcanic cones are covered with a young growth of

Pinus maritimus, with here and there a group of poplars, or of *Persea indica*. These, and particularly the first, are the trees which furnish the wood for the orange boxes, and on our way we saw several picturesque groups of bronzed, scantily-clad Açoreans, cutting down the trees, reducing the trunks to lengths suitable for the different parts of the boxes, and binding up the branches and unavailable pieces into scarcely less valuable faggots of firewood.

Every yard of tolerably level ground was under crop; maize chiefly, with here and there a little wheat, or a patch of potatoes or of tomatoes, or more rarely of sweet-potatoes, for here *Convolvulus batatas* seems to have nearly reached its temperature limit. Many fields, or rather patches—for each crop usually covers a small space which is not separated from the contiguous patches by any fence—are fallow; that is to say, are under a luxuriant crop of lupin, sown to be dug down bodily as manure, so soon as the plant shall have extracted the maximum of assimilable matter from the water and air.

After passing Ribeira Grande the road becomes more rugged, now passing down into a deep gorge with a little hamlet nestling in it, and a bridge spanning the dry bed of a wet-season torrent; and now rising over the well-cultivated spur of a mountain ridge.

We stopped for luncheon in a pretty little ravine, well shaded by trees and watered by a considerable stream.

Posting round the world as we are doing with very little spare time at our disposal, one impression succeeds another so rapidly, that it is sometimes not very easy to disentangle them in one's memory, and refer each picture to its proper place. This little valley, now ringing with English 'chaff' and laughter, and littered with the inevitable sardine-tins and soda-water bottles, seemed a reflex of our confused cosmopolitan condition of mind. The tall, smooth tree-boles, with their scanty blue aromatic foliage, all around us,—which made up the greater part of the vegetation,—were the gum-tree (*Eucalyptus robustus*), from New Holland. The group of beautiful dark conifers on the other side of the stream, showing in every tone of colour and in every curve of their long drooping branches their thorough luxuriance and ' at-homeness,' were no Atlantic or European cypresses, but *Cryptomeria japonica*, the lawn tree which saddens us with its blighted brown twigs after a too hard frost in England. The tree above it with the dark green phyllodes was *Acacia melanoxylon* from Australia; the livelier intermixed greens were due to the Japanese *Pittosporum undulatum*, to *Persea indica*, and *Laurus canariensis—* both of somewhat doubtful origin though reputed natives—and to the undoubtedly native *Myrica faya*.

The Açores have been particularly fortunate in having their climate made the most of by the introduction of suitable and valuable plants. When the islands were first discovered they were clothed with natural forest, but during the earlier period of their occupation the wood was cut down with so little judgment that it was almost exterminated, and it became necessary to send planks for orange-boxes from Portugal. Of late years, however, several of the wealthiest and most influential proprietors, both in Fayal and San Miguel, have interested themselves greatly in forestry and acclimatisation, and have scattered any of their new introductions which seemed to be of practical value about the islands with the utmost liberality. All the trees from Europe and the temperate parts of America, north and south, and those of Australia, New Zealand, Japan and the cooler parts of China, seed freely in the Açores, so that there seems to be no limit to their muliplication. A quick-growing wood is, of course, the great desideratum, as it is chiefly wanted for the building of fires, and of the scarcely less ephemeral orange-cases. For this latter purpose, *Cryptomeria japonica*, several species of *Eucalyptus*, *Populus nigra* and *angulata*, and *Acacia melanoxylon* are already supplanting *Pinus maritimus*, *Persea indica*, and *Laurus canariensis*.

A few miles further on, the road left the coast, and began to ascend so rapidly that, until we gained the top of the ridge, we had little help from our carriages and 'mulos.' The uplands, in general character and in the style of their vegetation, are not very unlike some of the richer parts of the High-lands of Scotland. The plants are somewhat on a larger scale. The 'heather' is the *Erica azorica*, frequently rising to the height of twelve to fifteen feet, with a regular woody stem much used for fire-wood. The bog-myrtle is replaced by the graceful *Myrica faya*, and the juniper is represented by a luxuriant spreading prostrate form, *Juniperus oxycedrus*. Grasses are numerous in species, and form a rich green permanent pasture. Ferns are very

abundant, and give quite a character to the vegetation of the ravines among the ' Montas.' The steep
cliff down to the bed of a torrent is sometimes one
continuous sheet of the drooping fronds of *Woodwardia radicans,* often six or eight feet in length.

The *Woodwardia* is certainly the handsomest and
most characteristic of these investing ferns. In the
glades in San Miguel it is usually associated with
the scarcely less handsome *Pteris arguta,* and with
many varieties of *Aspidium dilatatum* and *æmulum.*
Here and there we come upon a fine plant of
Dicksonia culcita, the nearest approach on the island
to a tree-fern. The buds and young fronds of this
fern are thickly covered with a soft, silky down,
which is greatly used in the islands for stuffing beds
and pillows.

On reaching the crest of the hill the view is certainly very striking. You find that you are on the
top of the ridge bounding an old crater of great
extent. The valley of the Furnas, richly cultivated
and wooded, lies directly below ; with a scattered
town, with public gardens, baths, and lodging-
houses, as an object of central interest. The valley,
at a first glance, looks strangely familiar from its
resemblance to many of the valleys in Switzerland.
It is not until the eye has wandered over the lava
ridges and rested upon the dense columns of vapour
rising from the boiling springs that one realises the
critical condition of things—the fact that he is
descending into the crater of a volcano, which still
gives unmistakable signs of activity.

The road into the valley is very steep, zigzagging
through deep cuttings down the face of the mountain.

It was about five o'clock when our now somewhat
weary cavalcade drew up before the door of the hotel
in the village.

We had been told by the British Consul at Ponta
Delgada that about four miles beyond the village,
following a bridle path across a ridge and along the
border of a lake, we should find a comfortable, commodious hotel, kept by an Englishman, where, if we
gave due notice, we could get all accommodation.
Unfortunately there was no time to give notice, so we
determined to go on chance.

One or two of us started off on foot, while the gear
was being transferred from the carriages to a train of
donkeys, to give Mr. and Mrs. Brown what preparation
we might, and to organize some dinner. We had a
lovely walk,—up a winding path among the rocks to
the top of a saddle, where a beautiful blue lake about
a couple of miles in length, bordered with richly-
wooded cliffs, lay below us. On the opposite side,
about a couple of hundred feet above the lake, we
could see Gren'a, Mr. Brown's house; and nearer us,
on the shore of the lake, a group of natural cauldrons,
where the water was bubbling and steaming, and
spreading widely through the air a slight and not
unpleasant odour of sulphur. No human habitation
except Mr. Brown's was visible; but though the scene
seemed singularly quiet and remote, its richness and
infinite variety in light and shade and colouring prevented any oppressive effect of extreme loneliness.

Mr. Brown met us at the door; we told him that
there were about a dozen of us who wanted rooms
and food, and he naturally answered that he had
nothing to give us, and put it to our common sense

how it could be possible that he, in his primæval solitude, should be ready at any moment to entertain a dozen hungry strangers, to say nothing of their servants and their asses. Notwithstanding, there was a reassuring twinkle in Mr. Brown's shrewd, pleasant eyes. We wrung an admission from him that there was plenty of room in the house, that fowls might be got, and eggs and tea. Mrs. Brown joined us, and her appearance was also reassuring; so we shouted for the urgent tub, and left the rest to fate. Shortly we saw the long string of asses winding, with our changes of raiment, round the end of the lake, and it was not to our surprise that about eight o'clock we found ourselves sitting before an admirable dinner, with all our arrangements for the next couple of days settled in the most satisfactory way. We sent the carriages back to Ponta Delgada, with orders to meet us at mid-day on Monday at Villa Franca, a town on the southern coast of the island; and we engaged some fifteen or twenty donkeys for Monday morning to take us and our effects over the ridge and down the steep passes to the shore road.

Next morning some of our party walked to the Roman Catholic chapel in the village, and afterwards went to see the hot springs; others wandered about on the slopes and terraces overlooking the lake, enjoying the quietude and beauty of the place.

But for the birds, which were numerous, and the distant murmur of the boiling springs, the silence was absolute. Now and then a large buzzard, *Buteo vulgaris*, on account of whose abundance the islands were first named from the Portuguese word *açor*, a kite, rose slowly and soared in the still air. A genuine

blackbird, *Turdus merula*, poised himself on the top
of a fir-tree and sang to us about home; a chaffinch,
Fringilla tintillon, very nearly genuine, hopped on the
path and acted otherwise much like an English chaffinch; a bullfinch, *Pyrrhula murina*, so like the real
thing as to have given rise to some discussion, piped
in the thicket; and the canary, *Serinus canarius*, here
no albino prisoner, but a yellow-green sparrow of unlimited rapacity in the way of garden-seeds, settled
on the trees and twittered in large flocks. I walked
down to the baths by a short cut across the hills with
Mr. Brown in the afternoon, and got a great deal of
pleasant information from him. It seems that he
was very much identified with the late rapid progress
of gardening and forestry. Between twenty and
thirty years ago he went from England, a young gar-
dener, to lay out the splendid grounds of M. José do
Canto at Ponta Delgada; he assisted in various
schemes of horticulture in the interest of M. Ernest
do Canto, M. Antonio Borges, and other wealthy proprietors, and among other things designed the pretty
little public garden at Furnas, which we passed through
on our way to the springs. The house which Mr.
Brown now occupies, with about four hundred acres
of land, belongs, singularly enough, to a London
physician, and Mr. Brown acts as his factor. It is
most comfortable and pleasant—just one of those
places to suggest the illusory idea of going back
sometime and enjoying a month or two of *rest*.

The principal boiling springs are about half a mile
from the village. Round them, over an area of
perhaps a quarter of a mile square there are scorched-
looking heaps like those which one sees about an

iron-work, only whitish usually, and often yellow from
an incrustation of sulphur. Over the ground among
one's feet little pools of water collect everywhere, and
these are all boiling briskly. This boiling is due,
however, chiefly to the escape of carbonic acid, and
of vapour formed below, for the temperature even of
the hottest springs does not seem to rise to above
90° C. The largest of the springs is a well about
twelve feet in diameter, inclosed within a circular
wall. The water hisses up in a wide column nearly
at the boiling point, bubbling in the centre to a height
of a couple of feet, and sending up columns of steam
with a slight sulphurous smell. A little further on
there is a smaller spring in even more violent ebulli-
tion, tossing up a column five or six feet high; and
beyond this a vent opening into a kind of cavern, not
inaptly called 'Bocco do Inferno,' which sends out
water, loaded with grey mud, with a loud rumbling
noise. The mud comes splashing out for a time
almost uniformly, and with little commotion, and
then, as if it had been gathering force, a jet is driven
out with a kind of explosion to a distance of several
yards. This spring, like all the others, is surrounded
by mounds of silicious sinter, and of lime and alumina
and sulphur efflorescence. The mud is deposited
from the water on the surface of the rock around in
a smooth paste, which has a high character all round
as a cure for all skin complaints. When I looked at
it first I could not account for the grooves running
in stripes all over the face of the rocks; but I afterwards found that they were the marks of fingers
collecting the mud, and I was told that such marks
were more numerous on Sunday, when the country

people came into the village to mass, than on any other day.

At a short distance from the 'Caldeiros' a spring gushes out from a crack in the rock of a cool chalybeate water, charged with carbonic acid, and with a slight dash of sulphuretted hydrogen.   There is a hot spring close beside it, and on the bank of the warm stream and in the steam of the Caldeira, there is a luxuriant patch of what the people there call 'ignami' or yams (*Caladium esculentum*), which seems to thrive specially well in such situations.   The flavour of the aërated water is rather peculiar at first, but in the hot steamy sulphurous air one soon comes to like its coolness and freshness, and it seems to taste all the better from the green cup extemporised out of the beautiful leaf of the *Caladium.*   The warm water from all the springs finds its way by various channels to join the river Quente, which escapes out of the 'valley of the caves' at its north-eastern end, and, brawling down through a pretty wooded gorge, joins the sea on the north coast about six miles from Villa Franca.

We left Gren'a after breakfast next morning, our long train of about twenty saddle and baggage asses winding along the eastern shore of the lake and up the steep passes—gloriously fringed and mantled with *Woodwardia* and *Pteris arguta*, and variegated with copses of the dark tree-heath and brakes of the bright green faya—to the crest of the ridge bounding the northern end of the valley ; and thence down crooked and laborious ways through many gorges planted with grafted fruit-bearing chestnuts, and over many lava spurs, to the road along the south

shore where we found the carriages waiting for us.
The wheat harvest was going on vigorously in the
lower lands, and shortly before entering Villa Franca,
a long town which straggles over four or five miles
between Ribeira Quente and Ponta Delgada, we
stopped and rested at a farm-house where they were
' threshing.' The carriage I was in had fallen a
little behind the rest, and when we came up the scene
at the farmyard was very lively. Outside was the
' threshing-floor,' a hardened round area with a stake
in the centre. The wheat was spread on the baked
clay floor, and two sledges, each drawn by a pair of
oxen, went slowly round and round, ' treading out
the corn.' The sledges were driven with much noise
and gesticulation by tawny, good-natured Açoreans,
and were often weighted by a mother or aunt squat-
ting on the sledge, holding a laughing, black-eyed
baby. The drivers were armed with enormously long
poles, with which they extorted a certain amount of
attention to their wishes from the unmuzzled oxen,
much more intent upon snuffling among the sweet
straw for the grains of wheat, and making the most
of their brief opportunity. Within the house whither
most of our party had retreated from the roasting
sun, the first large entrance room was encumbered
with the beautiful ripe ears of maize, of all colours,
from the purest silvery white to deep orange and red.
It was high noon however, and a lot of bright-eyed
girls, who had been husking the maize, had knocked
off work; and on the arrival of the strangers, a lad
brought out a guitar, and they got up a dance, very
simple and merry, and perfectly decorous.
Neither hosts nor guests understood one word

of the others' language, but by dint of signs, and laughter, and human sympathy generally, we got on wonderfully well. It seemed to be the steading of a well-to-do farmer. There were other houses in the neighbourhood, and a number of young people seemed to have congregated, so that we had a good opportunity of seeing some of the peasants. The men are generally good-looking, with spare, lithe, bronzed figures, dark eyes, and wide, laughing mouths, with fine white teeth. The women in the Açores are usually inferior to the men in appearance, but at this farm some of the girls were very good-looking also, with clear complexions, and more of a Spanish than a Portuguese type.

From Villa Franca we drove along the shore to Ponta Delgada, where we arrived early in the evening.

While we were at the Furnas some of our companions started in the other direction, to the Caldeira of the Sete-Cidades, and were greatly pleased with their trip. This crater is probably the most striking feature in the scenery of the island. The road to it from Ponta Delgada goes westward for some miles along the southern coast. It then gradually winds upwards through ravines festooned with *Woodwardia*, and among rugged, volcanic masses clothed with ' faya ' and tree-heath, to the top of a crest, between two and three thousand feet in height. A wonderful scene then bursts upon the wanderer. The ridge is the edge of a large crater two miles and a half in diameter, surrounded by an unbroken craggy wall, more than a thousand feet in height. The floor of the crater is richly wooded and cultivated. There

are two small lakes of a wonderful sapphire blue, and
on the margin of one of them a village of white
cottages. The zigzag path down into the crater is so
steep that one or two of the parties who went from
the ship contented themselves with the view of the
valley from the crest of the ridge, and from all I hear
I am inclined to think that these had the advantage
in every respect over some others, who went down
and had to come up again.

Next morning Captain Nares and I called on
M. José do Canto, about whose good and liberal
deeds in introducing valuable and ornamental
foreign plants and distributing them through the
islands, we had heard so much. We were fortunate
in finding him at home, and we spent a very pleasant
couple of hours with him in his charming garden.

The trees of all temperate and sub-tropical regions
seem to thrive admirably in sheltered situations in
the Açores. M. do Canto has for the last thirty
years spared neither money nor time in bringing
together all that appeared desirable, whether for
their use or for their beauty, and in doing them
ample justice while under his charge. The garden is
well situated on the slope above the town; it is
extensive, and very beautifully laid out and cared
for. Great care is taken to allow each individual
tree to attain its characteristic form, and consequently
some species, particularly those of peculiar and
symmetrical growth, such as the different species of
Altingia, Araucaria, Cryptomeria, &c., are more
perfect probably than they are anywhere else, even
in their native regions. M. do Canto does not give
much heed to the growing of flowers. His grounds

are rather an arboretum than a garden; he has now upwards of a thousand species of trees under cultivation.

We left Ponta Delgada on the 9th of July, and just before our departure we had an opportunity of seeing a singular religious ceremony.

In one of the churches of the town there is an image of our Saviour, which is regarded with extreme devotion. The inhabitants, in cases of difficulty or danger, bring it rich offerings, and the wealth of the image in jewels was variously stated to us at from 1,000*l.* to 100,000*l.*, in proportion to the faith and piety of our informants. There had been great want of rain in the island for some months past, and it had been determined to take a step which is taken only in extreme cases,—to parade the image round the town in solemn procession.

People began to come in from the country by mid-day, and all afternoon the town wore a gala appearance. The Açorean girls, as soon as they can afford it, purchase, if they have not already inherited a long, full, blue cloth cloak, coming down to the heels, and terminating in an enormous hood, which projects, when it is pulled forward, a foot at least before the face. The cloak and hood are thus a complete disguise, for if the lower part of the hood be held together by the hand, a very common attitude, while the eyes can be used with perfect freedom, both figure and face are entirely hidden. These cloaks and hoods are very heavy and close, and it seems strange that such a fashion can hold its ground where the conditions are very similar to those in the extreme south of Spain or Italy. The

head-dress of the men is singular, but it has a more
rational relation to the exigencies of the climate.   It
is also made of dark blue cloth,—a round cap with a
long projecting peak, and a deep curtain falling over
the neck and shoulders, an excellent defence whether
from rain or sun.   The odd thing about it is that
where the hat is made in the extreme of a bygone
'mode' which still lingers in the remote parts of the
island, the sides of the peak are carried up on each
side of the head into long curved points, like horns.
The horns are 'going out,' however, although a
general festa, such as we were fortunate enough to
see, still brought many grotesque pairs of them into
the city.

We saw the procession from the windows of the
principal hotel, which looked across a square to the
church from which it took its departure.   The square
and the streets below us were, for hours before, one
sea of carapuças and capotes, male and female, but
chiefly the latter, their wearers sitting on the hot
pavement, chattering quietly.   About five o'clock a
large number of acolytes in scarlet tunics left the
church, and formed a double row, lining the streets
in the path of the procession.   Then came a long
double row of priests in violet chasubles and stoles,
repeating the responses to a portly brother, who led
the column intoning from his breviary.   Then a
double row of priests in white, and then a group of
the higher clergy in cloth of gold and richly
'appareled' vestments, preceding the image, which
was carried aloft under a crimson canopy.   The
image was certainly not a high work of art, but
it seemed to be loaded with valuable ornaments.

Behind the canopy walked the civil governor Count de Praya de Victoria, the military governor, and some of the high state functionaries, and the procession was closed by a column of monks. As the image approached, the people knelt everywhere within sight of it, and remained kneeling until it was past. It is, of course, difficult for us to realise the convictions and feelings under which the inhabitants of San Miguel unite in these singular pageants. No one could doubt that the devotional feeling was perfectly sincere; and it was moderate, with no appearance either of gloom or of excitement; the manner of the large crowd was throughout grave and decorous.

We looked with great interest the next morning to see whether our friends had got the coveted rain, but although the peaks and ridges fringing the crater-valleys were shrouded under a canopy of cloud and mist, the sky looked as hard as ever, not a whit nearer the point of precipitation.

Our first haul after leaving Ponta Delgada was in 1,000 fathoms, midway between the islands of San Miguel and Santa Maria, and about fifteen miles north-west of the Formigas. The bottom was Globigerina ooze. The principal feature in this dredging was the unusual abundance of stony corals of the deep-sea group.

Three living specimens of a large species of *Flabellum* (Fig. 12) were sifted out, the same as the one which we had dredged previously at Station 73, to the west of Fayal. The corallum is wedge-shaped, the calicle rising from an attenuated pedicle. The extreme height, from the end of the pedicle to the

PLATE XVIII. *Meteorological Obs*

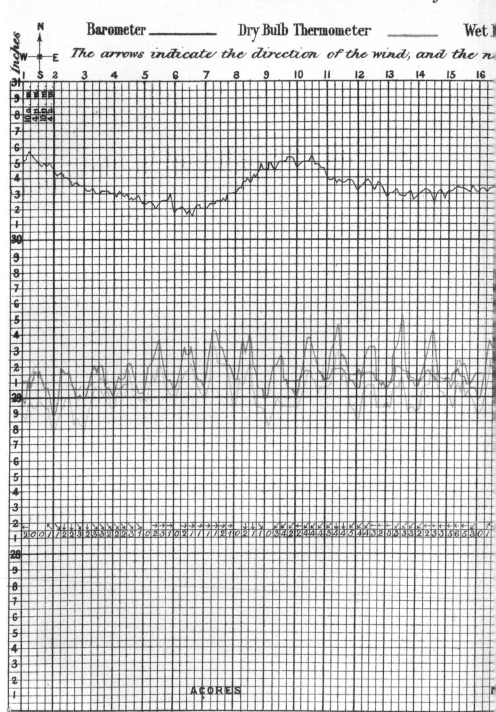

ACORES

vations for the month of July, 1873.

lb Thermometer   Temperature of Sea Surface

nbers beneath its force according to Beaufort's scale

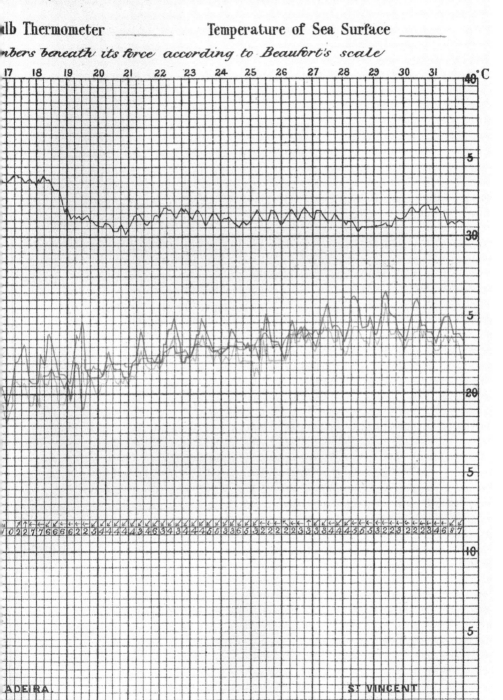

margin of the cup, is 50 mm. ; the greatest diameter of the calicle is 65 mm., and the smallest 30 mm. The three specimens are very nearly of the same dimensions.

The lateral costæ make an angle with one another of 120° to 140°, and are sharp and moderately prominent, with an irregular edge. The external surface of the calicle is covered with a glistering epitheca, and near the margin is of a light pink colour. The costæ of the faces corresponding to the primary and

FIG. 12.—*Flabellum alabastrum*, MOSELEY. Slightly enlarged. (No. 78.)

secondary septa are almost as well marked as the lateral costæ, and appear as irregularly dentated ridges, separated by slight depressions. The ends of the calicle are broadly rounded, and it is compressed laterally in the centre. The upper margin is curved, describing about one-third of a circle.

There are six systems of septa disposed in five cycles. The septa are extremely thin and fragile.

E 2

They are tinged with pink, and covered with rounded granules, disposed in rows. The primary septa are approximately equal to the secondary, giving somewhat the appearance of twelve systems. These septa are broad and prominent, with a rounded superior margin, and curved lines of growth. The septa of the third, fourth, and fifth cycles successively, diminish in breadth, and are thus very markedly distinguished from one another, and from the primary and secondary septa. The septa of the fourth cycle join those of the third a short distance before reaching the columella. The septa of the fifth cycle are incomplete. The margin of the calicle is very deeply indented, the costæ corresponding to the primary and secondary septa, being prolonged in conjunction with the outer margins of these septa, into prominent pointed processes; similar but shorter prolongations accompany the tertiary, and some of the quaternary septa. Between each of the sharp projections thus formed, the edge of the wall of the calicle presents a curved indentation.

Two of the specimens procured expanded their soft parts when placed in sea-water. The inner margin of the disc round the elongated oral aperture presents a regular series of dentations corresponding with the septa, and is of a dark madder colour; the remainder of the disc is pale pink. The tentacles take origin directly from the septa. They are elongated and conical. Those of the primary and secondary septa are equal in dimensions, and along with the tertiary tentacles, which are somewhat shorter, but in the same line, are placed nearest the mouth, and at an equal distance from it. The

tentacles of the fourth and fifth cycles are successively smaller and at successively greater distances
from the mouth.   Placed on either side of each
tentacle of the fifth cycle, and again somewhat
nearer the edge of the calicle, there are a pair of
very small tenacles, which have no septa developed in
correspondence with them.   There are thus four successive rows of tentacles, and the normal number is
ninety-six.   The tentacles are of a light red colour,
and between their bases are stripes of yellowish red
and light grey.

This form belongs to the group *Flabella sub-pedi-
cellata* of Milne-Edwards, and probably to that

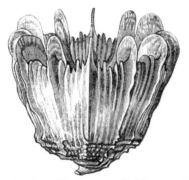

FIG. 13.—*Ceratotrochus nobilis*, MOSELEY.   Slightly enlarged.   (No. 78.)

division in which the costæ are prominent and ridge-
like on the faces of the corallum, as well as on its
lateral margins ; but it differs from those described
under this head by Milne-Edwards, in that it has
five cycles, the fifth being incomplete, and in other
particulars which appear from the description given.

A single living specimen of a coral, referred by Mr.
Moseley to the genus *Ceratotrochus* (Fig. 13), was

Temp.		Fms.
21°.5 C		Surf.
15°		80
14°		110
13°		150
12°		200
11°		250
10°		450
9°		600
8°		690
7°		740
6°		780
5°		850
4°		950
3°		1200
1°.8		2400

Fig. 14.—Diagram constructed from Serial Sounding (No. 82).

obtained from this haul. The orallum is white. The base sub-pedicellate, with a small scar of original adherence. The principal costæ are prominent, and round the region of the base they are beset with small spines directed somewhat upwards. The upper portion of the costa is without spines. The primary and secondary septa are broad and exsert. Pali are absent, the columella is fascicular. The absence of pali, the form of the columella, and the nature of the base, associate this form with the *Ceratotrochi* as defined by Milne-Edwards.

The animal is of a dark madder colour on the region of the margin of the calicle between the exsert primary and seconday septa, and on the membrane investing the wall of the corallum from the margin down to the commencement of the spines. This dark colour is succeeded on the disc by a band of pale blue, within which there is again a zone of very dark madder colour round the mouth. The

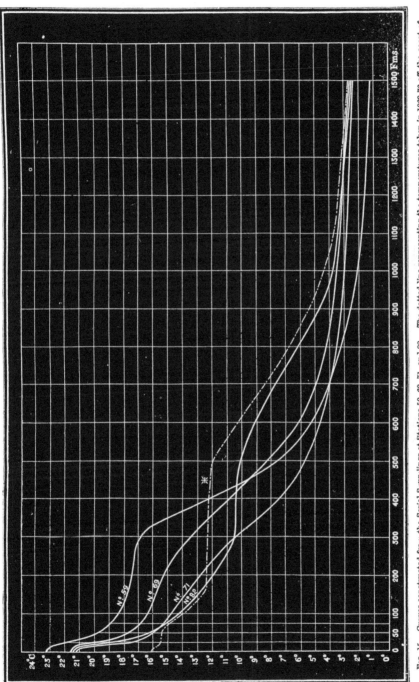

FIG. 15.—Curves constructed from the Serial Soundings at Stations 59, 69, 71, and 82.   The dotted line representing the temperatures taken in 1869-70 off the coast of Portugal is introduced for comparison.

dark colouring-matter is interesting, as it gives an absorption spectrum of three distinct bands.

On Friday, July 11, we sounded in 2,025 fathoms, 376 miles to the west of Madeira, the bottom very well marked 'globigerina ooze,' and the bottom temperature 1°·5 C.

On the following day the depth was 2,260 fathoms, the bottom Globigerina ooze, and the recorded bottom temperature 1°·8 C. ; and on the 13th the depth was 2,675 fathoms, with the same very characteristic Globigerina deposit, and a temperature of 2°·0 C. The bottom temperatures in this section show some irregularities, but as these do not extend beyond 0°·2 C., they may arise from errors of observation, due to the somewhat unsatisfactory mode of registering of Six's thermometers.

On the 14th we sounded in 2,400 fathoms; and a serial temperature-sounding (Fig. 14) indicated an almost total disappearance of the upper stratum of abnormally warm water ; but on the other hand the isotherms between three and eight hundred fathoms showed very distinctly the excess of heat in a deeper layer, to which reference has already been made, and which, becoming more marked a little to the northward, gives so peculiar a character to the temperature soundings in the Bay of Biscay. In Fig. 15 the curves constructed from the serial-soundings between Bermudas and Madeira show very clearly the gradual disappearance of the upper warm layer in passing to the eastward; and the appearance of the second deeper hump near the coast of Africa. The curve marked with the asterisk constructed from the *Porcupine, Lightning,* and *Shearwater* soundings is introduced for comparison.

The weather for the last few days had been remarkably fine, with a pleasant light breeze. When we turned up on deck on the morning of the 16th, we were already at anchor in the beautiful bay of Funchal, and looking at the lovely garden-like island, full of anticipations of a week's ramble among the peaks and 'currals' and the summer 'quintas' of our friends—anticipations in which we were destined to be disappointed.

SAN MIGUEL.

APPENDIX A.

Table of Temperatures observed between Bermudas and Madeira.

Depth in Fathoms.	Station No. 58. Lat. 32° 37' N. Long. 64° 21' W.	Station No. 59. Lat. 32° 54' N. Long. 63° 22' W.	Station No. 60. Lat. 34° 28' N. Long. 58° 56' W.	Station No. 61. Lat. 34° 54' N. Long. 56° 38' W.	Station No. 62. Lat. 35° 7' N. Long. 52° 32' W.	Station No. 64. Lat. 35° 35' N. Long. 50° 27' W.	Station No. 65. Lat. 36° 33' N. Long. 47° 58' W.
Surface.	23°·0 C.	23°·3 C.	22°·0 C.	21°·7 C.	21°·1 C.	23°·9 C.	22°·5 C.
25	...	...	...	19·5	...	...	...
50	...	...	...	18·9	...	...	...
75	...	...	...	18·3	...	...	...
100	...	18·4	17·8	17·7	17·4	17·3	17·6
200	...	17·4	17·0	17·2	16·8	16·8	17·4
300	...	16·9	14·9	15·6	14·3	14·8	16·1
400	...	12·2	11·0	12·0	11·2	11·3	13·9
500	...	7·7	6·8	7·9	7·2	7·6	11·0
600	...	5·3	5·0	...	5·0	...	7·2
700	...	4·1	4·1	...	4·3	...	5·3
800	...	3·7	3·9	...	3·7	...	4·7
900	...	3·3	...	...	3·5	...	4·2
1000	...	3·1	...	...	3·3	...	3·7
1100	...	3·0	...	...	3·0	...	3·3
1200	...	2·7	...	...	3·0	...	3·1
1300	...	2·8	...	...	...	...	2·9
1400	...	2·3	...	...	...	...	2·8
1500	2·3	2·3	...	...	...	...	2·7
Bottom Temperature.	2°·3	1°·7	1°·5	1°·5	1°·8	...	1°·7
Depth.	1500	2360	2575	2850	2875	...	2700

Depth in Fathoms.	Station No. 66. Lat. 57° 24' N. Long. 44° 14' W.	Station No. 67. Lat. 37° 54' N. Long. 41° 44' W.	Station No. 68. Lat. 38° 3' N. Long. 39° 19' W.	Station No. 69. Lat. 38° 23' N. Long. 37° 21' W.	Station No. 71. Lat. 38° 18' N. Long. 34° 48' W.	Station No. 72. Lat. 38° 34' N. Long. 32° 47' W.	Station No 73. Lat. 38° 30' N. Long 31° 14' W.
Surface.	21°·1C.	21°·1C.	21°·1C.	21°·7C.	21°·7C.	21°·7C.	20°·6C.
25	...	...	...	...	17·9	17·9	...
50	..	...	...	...	16·1	17·3	...
75	...	...	...	...	15·5	17·1	...
100	17·2	17·5	...	16·3	14·8	16·5	15·4
200	16·3	16·0	...	15·2	12·8	12·8	13·8
300	15·6	15·6	...	13·5	10·3	11·3	12·6
400	13·1	12·7	...	10·9	7·6	8·4	9·3
500	10·1	8·2	...	8·3	5·8	7·3	7·3
600	7·0	5·3	...	6·1	5·0	6·2	6·3
700	4·8	4·8	...	5·0	4·2	4·9	5·3
800	...	3·3	...	4·3	3·5	4·3	4·7
900	...	3·2	...	4·0	3·1	3·8	4·1
1000	...	3·2	...	3·7	3·0	3·1	3·7
1100	...	2·8	...	3·3	...	3·1	...
1200	...	2·8	...	3·1	...	...	...
1300	...	2·9	...	2·9	...	...	...
1400	...	2·8	...	2·8	...	...	...
1500	...	...	...	2·6	...	...	...
Bottom Temperature.	1°·8	1°·8	1°·6	1°·7	2°·2	2°·8	3°·7
Depth.	2750	2700	2175	2200	1675	1240	1000

Depth in Fathoms.	Station No. 76. Lat. 38° 11' N. Long. 27° 9' W.	Station No. 78. Lat. 37° 24' N. Long. 25° 13' W.	Station No. 79. Lat. 36° 21' N. Long. 23° 31' W.	Station No. 80. Lat. 35° 3' N. Long. 21° 25' W.	Station No. 81. Lat. 34° 11' N. Long. 19° 52' W.	Station No. 82. Lat. 33° 46' N. Long. 19° 17' W.	Station No. 83. Lat. 33° 13' N. Long. 18° 13' W.
Surface.	21°·1 C.	21°·7 C.	22°·0 C.	21°·7 C.	21°·7 C.	21°·5 C.	21°·1 C.
25	...	...	...	17·5	...	18·6	...
50	...	...	...	15·4	...	16·6	...
75	...	...	...	13·9	...	15·1	...
100	13·6	14·2	13·5	14·1	...	14·3	...
200	12·0	12·1	11·8	11·8	...	12·0	...
300	10·9	10·7	10·7	10·6	...	10·4	...
400	9·6	...	9·3	10·4	...	10·6	...
500	8·3	...	8·7	9·4	...	·9·6	...
600	7·2	...	7·5	8·7	...	...	...
700	5·3	...	6·7	7·0	...	7·9	...
800	4·8	...	5·4	6·0	...	5·7	...
900	4·2	...	4·1	4·4	...	4·8	...
1000	...	...	3·8	3·5	...	3·8	...
1100	...	...	3·4	...	...	3·5	...
1200	...	...	3·0	...	...	3·0	...
1300	...	...	2·6	...	...	3·2	...
1400	...	...	2·4	...	...	2·6	...
1500	...	...	2·2	...	...	2·5	...
Bottom Temperature.	4°·2	...	...	1°·8	2°·0	1°·8	2°·2
Depth.	900	1000	2025	2660	2675	2400	1650

APPENDIX B.

Table of Specific Gravities observed between Bermudas and Madeira.

Date 1873.	Lati- tude N.	Longi- tude W.	Depth of the Sea.	Depth (d) at which the water was taken.	Temperature (t) at d.	Temperature (t') during observation.	Specific Gravity at (t). Water at 4° = 1.	Specific Gravity at 15°·56. Water at 4° = 1.	Specific Gravity at (t). Water at 4° = 1.
			Fms.	Fms.					
June 14	32° 54′	63° 22′	2360	Surface.	23°·3C.	25°·4C.	1·02432	1·02726	1·02517
,,	...	...		Bottom.	1·6	24·6	1·02411	1·02660	1·02857
,, 15	33 41	61 28		Surface.	22·8	23·5	1·02498	1·02712	1·02515
16	34 28	58 56	2575	Surface.	21·7	23·0	1·02516	1·02715	1·02552
,,	...	...		Bottom.	1·5	24·0	1·02482	1·02711	1·02909
,, 17	34 54	56 38	2800	Surface.	21·7	22·8	1·02520	1·02713	1·02549
18	35 7	52 32	2875	Surface.	21·1	22·5	1·02587	1·02722	1·02576
,,	...	...		150	17·2	23·3	1·02506	1·02711	1·02670
,,	...	...		250	16·2	23·2	1·02488	1·02693	1·02677
,,	...	...		500	7·2	23·25	1·02409	1·02614	1·02768
,,				Bottom.	1·8	23·2	1·02510	1·02715	1·02912
,, 19	35 29	50 53	2750	Surface.	21·7	23·1	1·02524	1·02726	1·02562
,,	...	...		Bottom.	...	19·7	1·02512	1·02619	1·02817
,, 21	36 33	47 58	2700	Surface.	22·5	23·2	1·02522	1·02727	1·02541
,,				Bottom.	1·7	23·7	1·02384	1·02605	1·02804
,, 22	37 24	44 14	2750	Surface.	21·1	22·4	1·02536	1·02716	1·02570
,,	...	...		Bottom.	1·8	23·6	1·02413	1·02630	1·02828
,, 23	37 54	41 44	2700	Surface.	21·1	21·7	1·02542	1·02700	1·02555
,,	...	...		Bottom.	1·8	21·0	1·02478	1·02619	1·02817
,, 24	38 3	39 19	2175	Surface.	21·1	23·4	1·02483	1·02694	1·02481
,,	...	...		150	15·7	23·2	1·02482	1·02687	1·02680
,,	...	...		250	14·3	23·3	1·02443	1·02651	1·02676
,,	...	...		500	8·3	23·25	1·02404	1·02608	1·02741
,,				Bottom.	1·7	20·6	1·02487	1·02617	1·02815
,, 25	38 23	37 21	2200	Surface.	21·7	23·2	1·02513	1·02718	1·02511
26	38 25	35 50	1675	Surface.	21·8	22·1	1·02540	1·02714	1·02547
27	38 18	34 48	1675	Surface.	21·1	22·4	1·02520	1·02700	1·02555
,,				Bottom.	2·3	20·0	1·02557	1·02670	1·02865
,, 28	38 34	32 47	1240	Surface.	21·7	22·6	1·02536	1·02724	1·02560
29	37 47	31 2		Surface.	21·1	21·8	1·02539	1·02701	1·02556
30	38 30	31 14	1000	Bottom.	3·7	19·2	1·02601	1·02693	1·02882
July 3	38 11	27 9	900	Surface.	21·1	21·4	1·02550	1·02701	1·02556
,,	...	...		150	12·7	18·7	1·02585	1·02664	1·02728
,,				Bottom.	4·2	18·4	1·02619	1·02691	1·02877
,, 4	37 52	26 26	750	Surface.	20·9	21·5	1·02534	1·02688	1·02547
,,	...	...		Bottom.	...	20·8	1·02543	1·02679	1·02877
,, 12	35 3	21 25	2660	Surface.	21·7	22·2	1·02588	1·02714	1·02550
,,	...	...		600	8·7	18·4	1·02598	1·02669	1·02797
,,				Bottom.	1·8	20·5	1·02478	1·02604	1·02803
,, 13	34 11	19 52	2675	Surface.	22·0	22·7	1·02525	1·02715	1·02543
14	33 46	19 17	2400	Surface.	21·5	21·8	1·02555	1·02717	1·02560
,,				Bottom.	1·8	21·2	1·02552	1·02699	1·02897
,, 15	33 13	18 13	1650	Surface.	21·1	21·7	1·02585	1·02746	1·02601
,,	...	...		Bottom.	2·2	20·0	1·02517	1·02629	1·02827

CHAPTER II.

MADEIRA TO THE COAST OF BRAZIL.

Return to Madeira.—The Black Coral.—*Ophiacantha chelys.*—*Ophiomusium pulchellum.*—*Ceratias uranoscopus.*—The Island of San Vicente.—Porto Praya.—The Island of San Iago.—A Red-coral Fishery.—The Guinea Current.—*Balanoglossus.*—Luminosity of the Sea.—*Pyrocystis.*—Young Flounders.—*Bathycrinus aldrichianus.*—*Hyocrinus bethellianus.*—St. Paul's Rocks.—Fernando Noronha.—Low bottom temperatures under the Equator.—*Ceratotrochus diadema.*—*Pentacrinus maclearanus.*—Dredging at moderate depths.—Arrival at Bahia.

WHEN we reached Madeira we found, to our great regret, that shortly before our arrival there had been a rather severe epidemic of small-pox in the town; and as Captain Nares thought it imprudent to give general leave, our stay was greatly abridged. One or two of the officers went on shore and enjoyed a short ramble over the lovely island, now in the height of its summer beauty ; and a few of our friends visited us on board while we were taking in our stock of fresh provisions, and made our cabins gorgeous with offer-- ings of flowers and fruit.

We left Funchal on the evening of the 17th of July, and proceeded towards San Vicente in the Cape Verde group. We took a temperature-sounding on the 18th, and on the 19th sounded and dredged in 1,125 fathoms, with a bottom of volcanic sand, a few miles to the west of the island of Palma. The dredging was fairly successful, yielding one or two undescribed echinoderms. On the evening of the 20th we were approaching the position of Station 3, where we had brought up the coral coated with manganese on the 18th of February, and we were anxious to have another cast as nearly as possible on the same spot in the hope of perhaps getting some of the coral alive, or in some way clearing up the question of its conditions. On the following day we were a little too far to the westward, so we steamed up near the desired point and sounded again upon the ridge in 1,675 fathoms, and put over the dredge. The event showed that we were not far out of our reckoning, for the dredge brought up a quantity of fragments of the coral, and several other animals identical with those procured in the previous haul. None of the coral was alive, however, and the pieces were quite similar in every respect to those which we had got before, so that no further light was thrown upon the curious question of its occurring in that peculiar semi-mineralized state at so great a depth.

I give here a preliminary notice, under the name of

Ophiacantha chelys (Figs. 16 and 17), of a pretty little brittle-star which was found clinging to several of the branches of coral. It is, however, so different in aspect from such typical species of the genus as *O. spinulosa* and *O. setosa* that I have some hesitation in associating it with them ; indeed I should scarcely have done so had it not been that the described form which approaches it most nearly is

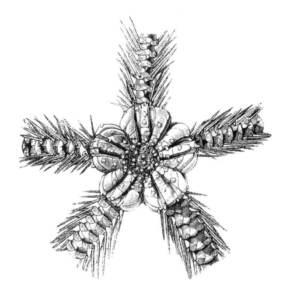

FIG. 16.—*Ophiacantha chelys*, WYVILLE THOMSON. Dorsal aspect of the disc. Four times the natural size. (No. 87.)

undoubtedly *Ophiacantha stellata*, LYMAN. I think it very likely that when we have an opportunity of studying the mass of new material which has been pouring in for the last three or four years, it may be found necessary to reconsider the genera of the Ophiuridea as at present defined, and to revise their limits. The diameter of the disc in *Ophiacantha chelys*

is, in an ordinary example, 8 mm.; the width of the
arm near the base 2 mm., and the arm is about
three-and-a-half times the diameter of the disc in
length. The disc is incised in the centre of the space
between the arms so deeply as to give it the effect of
being divided nearly to the centre into five broad
radial lobes; these lobes are a good deal inflated, and
each lobe is traversed in a radial direction by two
deep grooves, so that a deep outer rim of the upper

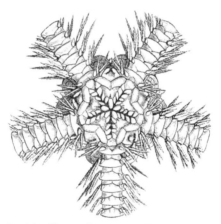

FIG. 17.—*Ophiacantha chelys*, WYVILLE THOMSON. Oral aspect of the disc. Four times the
natural size. (No. 87.)

surface of the disc is strongly fluted. The space in
the centre of the disc corresponding with the middle
third of its diameter is flat, and considerably depressed beneath the level of the outer inflated rim.
The whole of the surface of the disc is tessellated
with a certain approach to regularity with strong
calcareous plates, those towards the periphery larger
than those near the centre; and the plates bear small
stump-like spines, each with a crown of spinules on

the free end, inserted into distinct sockets hollowed out in the plates. The radial shields are long and narrow, and lie in the bottom of the grooves in the radial lobes; so that the shields of each pair are separated from one another by a high calcareous arch, almost a tube, formed of the inflated calcified perisom. The spines are specially congregated on the central depressed portion of the disc.

The mouth-papillæ are nine for each angle; they are broad and rather blunt, with the exception of the odd papillæ terminating the strong prominent jaws beneath the rows of teeth, which are larger than the others and pointed. There are no tooth-papillæ; the teeth are about five in a row, pointed and compressed vertically. The mouth-shields are large and wide, and rudely diamond-shaped; at the outer angle the sides of the plate are turned up a little, so as to form a short spout-like extension towards the base of the interbrachial groove. The first lower arm-plates are shield-shaped, the points closing the distal ends of the mouth-fissures; and those beyond are wide and crescentic, extending across the whole width of the arm. The tentacular scales are simple and leaf-like one to each tentacle. The side arm-plates are very large, meeting both above and below; they are raised distally into a high ridge marked with the shallow sockets of the arm-spines. The height of this ridge and the contraction of the proximal sides of these plates give the arms a peculiar beaded appearance. The arm-spines are seven in number on each side of two or three of the proximal arm-joints, and usually five on the joints towards the middle of the arm; they are long and glassy and

PLATE XIX. *The Track of the Ship from Madeira to Station 102.*

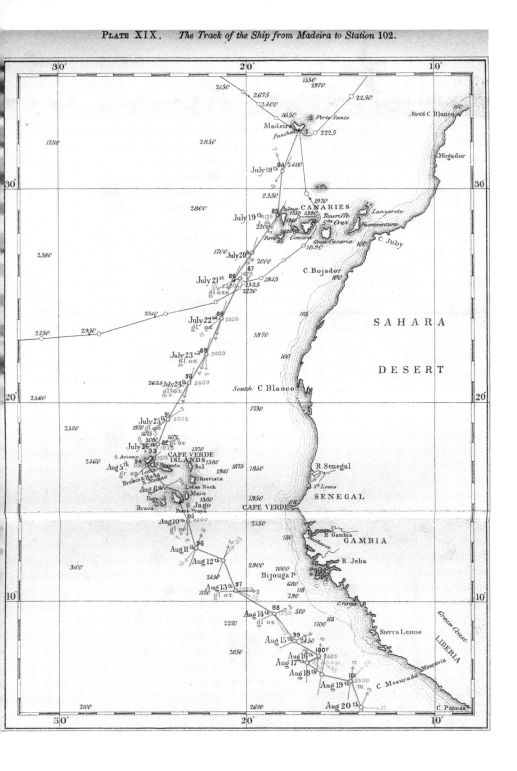

elegantly spinulated.   In colour the perisom passes
from a dark rose on the surface of the disc and along
the upper surface of the arms, through paler shades
to a nearly pure white on the under surface.

This haul yielded, along with *Ophiacantha chelys*,
the beautiful little Ophiurid represented in Figs.
18 and 19.   The diameter of the disc is 5 mm., and
the arms, which are rather wide at the base and taper

FIG. 18.—*Ophiomusium pulchellum*, WYVILLE THOMSON.   Dorsal aspect of the disc.   Seven
times the natural size.   (No. 87.)

rapidly, are only about once and a half the diameter
of the disc in length.   The upper surface of the disc
is very regularly paved with thick, well-defined plates,
each of which rises in the centre into a pointed
tubercle approaching a spine in character.   One
almost regularly hexagonal plate occupies the centre
encircled by a row of six plates of the same form ;

F 2

and beyond these there is an outer row consisting
of the five pairs of thick radial shields and five
oblong plates occupying the interbrachial spaces.
The mouth-papillæ are entirely coalesced into a
continuous calcareous border; the mouth-shields are

FIG. 19.—*Ophiomusium pulchellum*, WYVILLE THOMSON. Oral aspect of the disc. Seven times
the natural size. (No. 87.)

diamond-shaped and rather small; the side mouth-
shields, on the contrary, are unusually large. The
first four or five under arm-plates are shield-shaped
and rather large, with well-marked rounded tentacle-
scales; but they suddenly become small when they

reach the narrow part of the arm, and the tentacle-scales disappear. The side arm-plates on the proximal joints of the arms are very long—so long that those of one arm nearly meet those of the two adjacent arms, one or two small irregular plates only intervening; and the lower surface of the disc is thus made up to a great extent of the expanded bases of the arms. The side-plates on the distal arm-joints retain their unusual length, but they are directed outwards towards the end of the arm, and the inner edges of the plates of each pair are apposed throughout nearly their whole length both above and below.

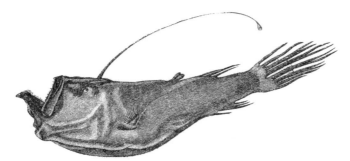

Fig. 20.—*Ceratias uranoscopus*, Murray. Natural size. (No. 89.)

The upper arm-plates are small and diamond-shaped; the arm-spines are of moderate size—usually three on each side arm-plate. I relegate this pretty little thing provisionally to the genus *Ophiomusium*, subject to reconsideration.

We sounded again and took temperatures on the 22nd, and on the 23rd we sent down the trawl to a depth of 2,400 fathoms with a bottom of 'globigerina-ooze;' along with a number of invertebrates this haul yielded a very singular little fish of the Lophioid family, which Mr. Murray has named *Ceratias urano-scopus* (Fig. 20). The specimen is 90 mm. in length from the snout to the end of the tail; compressed laterally and of a uniform black colour. The anterior spine of the first dorsal fin is produced into a long filament, ending in a pear-shaped bulb, terminating in a very distinct semi-transparent whitish spot. This spine has its origin on the posterior portion of the head, and when laid back it reaches nearly to the tip of the tail. The second part of the first dorsal is placed far back on the body, and consists of two short fleshy tubercles which lie in a depression in front of the second dorsal fin. The second dorsal has three rays; the anal is opposite the second dorsal and has four rays; the caudal has eight rays, the four central rays being much larger than the others and bifid. The pectorals are small, and have ten very delicate rays. The gill-opening is a slit situated below the pectoral fin. The upper jaw is formed by the intermaxillaries, and is armed, together with the lower jaw, with a series of teeth of moderate size which can be depressed inwards as in *Lophius*. The skin is thickly covered with minute imbedded conical spines. The eyes are very small, and are placed high up on the middle of the head. The presence of a fish of this group at so great a depth is of special interest. From its structure and from the analogy of its nearest allies there seems to be no reasonable doubt that it lives on the bottom. It is the habit of many of the family to lie hidden in the mud with the long dorsal filament and its terminal soft expansion exposed. It has been imagined that the expansion is used as a bait to allure its prey, but it seems more likely that it is a sense-organ intended to give notice of their approach.

On the two following days we went on our course towards San Vicente, sounding and taking serial temperatures daily. The weather was very fine with a light north-easterly breeze. The water which had previously been of a deep blue colour changed on the 23rd to a dull green; on the 25th it resumed its beautiful shade of cobalt. On the 26th we trawled in 1,975 fathoms with small result, and on the 27th we anchored off Porto Grande.

We remained a week at San Vicente. The island is most uninteresting; bare ridges of reddish volcanic lavas and tuffs—some of them certainly with a rugged and picturesque outline; and wide valleys and valley plains,—wildernesses of fragments of the rock which look and almost feel as if they were at a low red-heat. It was now about the hottest season, and everything was dried up and parched; the water-courses were dry and all the vegetation had disappeared, except the weird-looking succulent weeds of the desert, which with their uncouth wrinkled forms and venomous spines looked like vegetable demons that could defy the heat and live anywhere. Here and there outside the town, where the carcase of a dead bullock or a horse had been flung out on the shingle or only half buried in it, polluting the air far and near, there were half a dozen of the Egyptian vulture (*Neophron percnopterus*) perched lazily upon the bones, and when disturbed flying off slowly and alighting again at a distance of a few yards. A curious incident gave us a ghastly interest in the movements of these foul birds. A very excellent

seaman-schoolmaster, Mr. Adam Ebbels, whom we
had taken with us from England, died suddenly just
before we reached Bermudas, and his successor was
to have joined us at Porto Grande. He came out in
the same steamer with a sub-lieutenant who was also
going to join the ship. They arrived ten days before
the ' Challenger,' and the schoolmaster put up at the
French hotel. On the Sunday before our arrival he
went out to take a walk, and had not since been heard
of. Of course, besides taking all the necessary official
steps, we were all on the watch for traces of him,
and we were told that if he were dead the vultures
would be our surest guides to the place where the
body lay. They have rather an unusual mode of
looking at some things at San Vicente. When we
were making inquiries about the missing schoolmaster the general impression seemed to be that he
had met with foul play, as he was known to have had
a small sum of money about him and a rather
valuable watch when he left the hotel; and we were
told further that a murderer lived in a cottage at a
little distance from the town. It seems that there is
good reason to believe that this man, who had been
originally sent to San Vicente for the good of
Portugal, had made away with several people during
his stay on the island. Although his profession was
by no means spoken of with approval, it was talked
of easily and freely, and he did not appear to be
entirely beyond the pale of society. I had a curiosity
to meet a murderer without having the responsibility
of any fiscal relations with him, and made an
arrangement to call at his cottage; but something
came in the way and prevented the visit.

It turned out, however, that the poor fellow had
not been murdered or robbed at all. His body was
found a week or two after we left, lying, dried up
with the scorching heat, on a ledge near Wellington
Peak; he had wandered too far and had been over-
come by heat and fatigue and unable to return—very
probably he had had a sun-stroke. His purse and
watch were intact; even the vultures had failed to
discover him, he had gone too far beyond the
ring round the town where they chiefly find their
food.

Fresh water is about the most important element
at San Vicente, for although heavy rains fall now
and then, sometimes the island is for a whole year
without a shower. The water is taken from deep wells
sunk through the tufaceous rock, and as the supply
is limited, the wells are carefully enclosed and pro-
tected, and closed except at certain times. A large
well just behind the town, in an octagonal building
covered in with a low-pitched roof, is the great centre
of attraction; thither from early morning one can
see files of stately negresses marching with large
rather elegantly formed earthen vases poised upon
their heads; and it is amusing to watch the congregation of them good-naturedly helping one another
to draw the water and to fill their pitchers; and
chattering and laughing, and most generously ex-
hibiting their serviceable rows of pearly teeth. I
think the negroes at San Vicente are certainly better
looking than those in the West Indies. Their figures
are slighter and they have altogether a lighter
effect. No doubt this carrying of water-jars has a
great influence in producing the erect gait and

ease of gesture for which the women especially are remarkable.

Some of the wells outside the town are almost picturesque. The well-building is usually enclosed within a white-washed stone wall, and as there is a little moisture and shade within the enclosure, generally two or three trees of respectable dimensions rise over the wall. There is usually a latticed gate of entrance with an ever-changing group of lively good-natured beings, as black as Erebus, clustering round it.

We left Porto Grande on the 5th of August and proceeded on our course towards Porto Praya, the principal town on the large island of San Iago, and the seat of the central government of the Cape Verde group. On the 6th, the fine peak of the active volcano on the island of Fogo was in sight, and early on the morning of the 7th we anchored off Porto Praya. Although the anchorage is more exposed and not nearly so suitable for the habitual resort of shipping, San Iago has greatly the advantage of San Vicente ashore. The town of Porto Praya is tidy and well-ordered; the government and municipal buildings are commodious; and the central Praça is really ornamental, with a handsome fountain in the middle, and an encircling row of irrigated and cared-for trees. At one end of the town there is a fine public well. The water is led in closed pipes, from a stream coming down from the higher land, into a large stone-built reservoir; from which there is a daily distribution from a long range of ornamental basins and spouts to a constant crowd of applicants.

The country, although on the whole somewhat arid and bare, is much less so than San Vicente. There is a large grove of cocoa-nut trees behind the town ; some of the streams are permanent, and the valleys are consequently much greener, and in some places they are luxuriantly fertile. The day we arrived we rode to the pretty little village of Trinidad ; the first part of our way was very desolate, over an expanse of hot gravel relieved here and there by trailing gourds and convolvuluses, and a scrub of castor-oil plant and a low-growing almost leafless *Acacia*, with long, wicked, white spines. We passed two or three fine examples of the celebrated 'Baobab-tree' (*Adansonia gigantea*) ; the trunk of the largest on our route was about 50 feet in circumference, but in some trees of the same species on the neighbouring coast of Africa, which are supposed to be among the oldest trees in the world, they attain the enormous dimensions of 30 feet in diameter. The Baobab-tree with its spreading low crown and large pendulous greenish-purple flowers has a very striking and unusual appearance.

After riding a few miles we came suddenly to a sort of basin at the head of the valley, with a slow stream passing through it and a broad belt of the most luxuriant tropical vegetation on either side. Groves of cocoa-nuts extended for miles along the banks ; and the land was cleared and fenced for the cultivation of yams, sweet-potatoes, maize, pumpkins, and all the ordinary vegetable productions of the tropics. Wherever the vegetation was allowed to run wild, it passed into a tangled thicket of oranges, limes, acacias, and castor-oil shrubs ; the whole so warped

and felted with climbing gourds and beautifully
coloured *Ipomeas* that it was no easy matter to make
one's way through it unless by the cleared tracks.
We went a little way up the flank of one of the hills
to the village, and had a good view of the valley,
which contrasted wonderfully in its extreme richness
and careful cultivation with the arid plains below.
The swarthy inhabitants received us with their usual
good-natured hospitality, and after a welcome luncheon of which bananas, oranges, pine-apples, and
cocoa-nut milk formed the principal part, we rode
back to the ship highly pleased with our experience
of this unexpected oasis.

Next morning one or two of us went out in the
steam-pinnace to dredge for red coral. We had
learned that there was a regular coral fishery on the
coast of San Iago, seven or eight boats being constantly employed, and nearly a hundred men; and
that coral to the amount of upwards of 100 quintals
(10,000 kilos.) was exported annually. The fishery
is carried on at depths between 60 and 100 fathoms
a mile or so from the shore. Large clumsy fishing-
boats are used, with a crew of from six to eight or
nine men. A frame of two crossed bars of iron,
weighted in the centre with a large stone, and hung
with abundant tangles, some of them of loose hemp
and others of net, is let down with a thick rope (one
and a half or two inch), and eased back and forward
on the ground till it has fairly caught; the rope is
then led to a rude windlass in the middle of the boat,
and it often takes the whole strength of the crew to
bring the frame up. The branches of coral stick in
the tangles and in the meshes of the net. It was a

fearfully hot day, the hottest I think in its physio-
logical effect on the human body which I have ever
experienced.   There was not a breath of air, and the
sea was as smooth as glass; and the vertical sun and
the glare from the water were overpowering.   We
crouched, half sick, under our awning, muffled up
to prevent the skin being peeled off; and even a
few successful hauls in the afternoon, which yielded
perhaps twenty or thirty fair branches of coral,
scarcely restored our equanimity.   A few of our first
hauls were unsuccessful, so we steamed up close to
one of the nearest fishing-boats.   The coral-fishers,
having no fear of competition, were very civil; indicating by signs when we were on the right spot.
They were active, swarthy Spaniards, and had stripped
themselves for their work to a pair of very scanty
drawers, and their lithe bronzed figures heaving round
the windlass were most picturesque; they got several
pieces of coral while we were out.   According to our
experience the coral grows at Porto Praya in loosely
spreading branches, from two to perhaps eight inches
high, attached firmly to ledges of rock and large
stones.   It is bad dredging-ground; our dredge got
jammed more than once, and was extricated with
difficulty.   The Cape Verde coral is not of fine
quality; it is dark and coarse in colour, and it
does not seem to be so compact in texture as the
Mediterranean variety.

The next morning a large party started on horseback in the direction of San Domingo.   We rode
over some hot flat country covered with a brush of
Acacia and *Ricinus,* and at length reached a ravine
with a small stream running in the bottom of it, the

banks fairly wooded, the wood interrupted every here
and there with spaces of loose stones and gravel. As
we rode along we frequently heard the harsh cry of
the Guinea-fowl, and Captain Maclear and I detached
ourselves from the riding party and spent most of
the day stalking a flock of them. They were
very wary, running very quickly, and rising and
taking a short flight before we could get within the
longest range. They crouched and ran among the
stones, and their speckled plumage so closely resem-
bled at a distance the lichen-speckled rocks, that
more than once when we had seen them moving
about, and had crept up within shot thinking that
we had kept our game constantly in sight, there was
nothing there but a heap of grey stones. In the
afternoon Captain Maclear managed to separate
some of the birds from the flock, and marked one
for his own ; he stalked it warily along the rugged
bank, and at last circumvented it, and cautiously
brought up his gun. A sharp report, and the fowl
fell. But Maclear's conscience was not to be burdened with the death of that beautiful, and, I may
add, delicious bird. At that moment a laugh of
triumph rang from the other side of a low ridge, and
Captain Nares, who, quite unconscious of our presence, had been stalking another flock in the same
direction, ran up and stuffed it into his game-bag.
Maclear had driven his bird right up to the muzzle of
Nares's gun ! I did not get a shot at a Guinea-fowl
either all day, but I picked up a few birds, and I
found the pretty 'king-hunter,' *Dacelo jagoensis*,
sitting tamely on the tops of the castor-oil bushes
where Darwin left him forty years before.

On the 9th of August we weighed anchor, and
proceeded on our course towards Fernando Noronha.
The northern limit of the equatorial current, running
westward at the rate of from twenty to seventy miles
a day, is, roughly, the fourth degree of north latitude ;
—a little to the southward of this parallel towards
the coast of Africa, considerably to the northward
about 35° W. longitude, where it approaches its
bifurcation off Cape San Roque.

Occupying a band approximately between the
parallels of 4° and 8° N., there is a tolerably constant
current to the eastward, the 'equatorial counter-
current,' averaging, in the summer and autumn
months when it attains its maximum, a rate of
twenty to forty miles a day. The causes of this
current are not well known ; it occupies a portion
of the ever-varying space between the north-east and
the south-east trades, and it seems probable that it
may be a current induced in an opposite direction, in
the 'zone of calms,' by the rapid removal of surface-
water to the westward by the permanent easterly
wind-belts. Opposite Cape Verde this easterly current takes a southward direction, it is joined by a
portion of the southern reflux of the Gulf Stream ;
and, under the name of the 'Guinea Current,' courses
along the African coast as far south as the Bights
of Benin and Biafra, where it disappears.

The 'Guinea' or 'African' current is a stream of
warm water, averaging from 250 to 300 miles in
width, with an average rate of from twenty to fifty
miles a day. Its greatest concentration and force is
opposite Cape Palmas, where it is jammed in by the
northern edge of the equatorial current ; its width is

there reduced to a little over a hundred miles, and it attains a maximum speed of one hundred miles a day. There seems to be no doubt that this current must be regarded as a continuation of, and as being almost entirely derived from, the equatorial counter-current. It is evident that a great part of the surface-water must have an equatorial origin, for when we took our observations, nearly at the hottest time of the year, the surface-temperature was equal to the mean maximum temperature of the air, and one degree above its mean minimum temperature ; it is doubtless joined, and considerably augmented, by a cooler current passing down the coast of Africa, past the Canary and Cape Verde Islands, a portion of the southern branch of the Gulf Stream ; and this tributary stream, whose direction so nearly coincides with that of the Guinea Current, formerly tended to prevent the full recognition of the principal source of the latter in the equatorial counter-current.

After leaving San Iago, on the 9th of August, we began almost at once to feel the influence of the Guinea Current, or rather perhaps of its northern tributary ; and from that date to the 17th our course lay in a south-easterly direction, parallel with the coast of Africa, and nearly in the path of the current. The temperature of the sea-surface during this time was nearly constant at 26° C., and the temperature of the air slightly lower. Serial soundings were taken at several stations, and these gave a singularly rapid fall in temperature of from 14° to 15° C. for the first 100 fathoms ; showing that the warm current, as in all other cases which we have observed, is very superficial. Where the rate of the current is highest, we

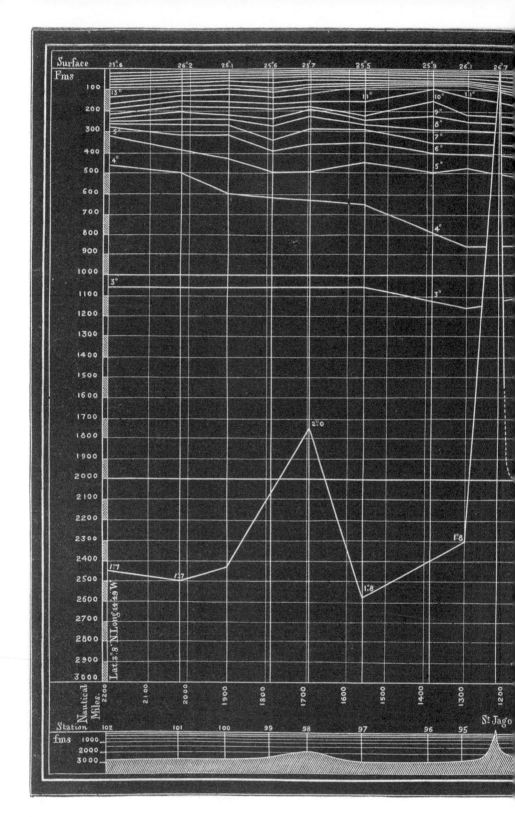

PLATE XX.—DIAGRAM OF THE VERTICAL DISTRIBUTION

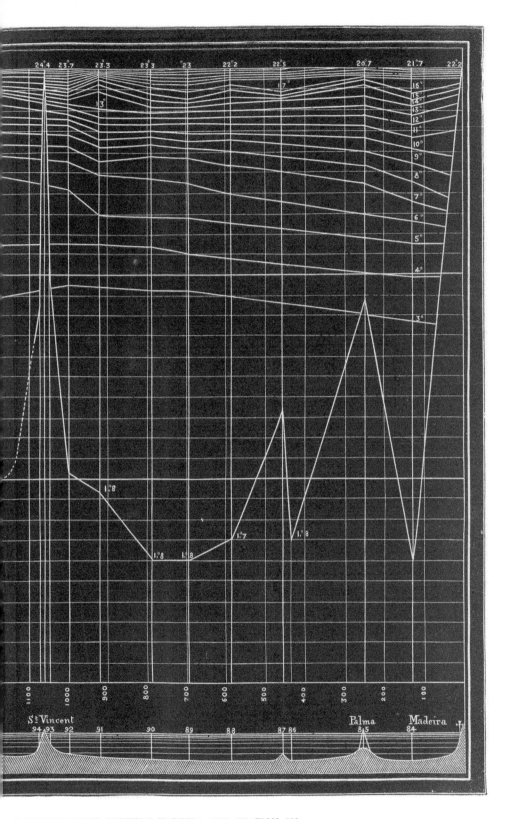

OF TEMPERATURE BETWEEN MADEIRA AND STATION 102.

have as usual the isotherms crowding upwards; the cooler water rising to supply the place of the hot surface-water, which is being rapidly drifted and evaporated away.

We sounded on the 10th in 2,300 fathoms with a bottom of 'globigerina-ooze,' and took a series of temperatures at intervals of 100 fathoms, down to 1,500. The surface-temperature was high, and from the surface the temperature fell with unusual rapidity, losing nearly 15° C. in the first hundred fathoms.

Surface	26°· 1 C.	15 fathoms . .	20 · 5 C.
5 fathoms . . .	25 · 4	20 ,, . . .	18 · 4
10 ,, . . .	24 · 4	100 ,, . . .	11 · 3

There was a marked tendency at this station to the gathering together upwards of all the higher lines (Pl. XIX.), the isotherm of 6° C. occurring at a depth of four hundred fathoms, nearly four hundred fathoms higher than the position of the same line at Madeira.

The following day we again took a series of temperature observations, and the gathering upwards of the warmer lines was still more marked (Station 96); and on the 13th a series of observations at intervals of 100 fathoms to a depth of 1,500 gave a like result. The fall of temperature for the first hundred fathoms was much the same as on the 10th.

Surface	25°· 5 C.	75 fathoms . .	12°· 7 C.
25 fathoms . . .	20 · 6	100 ,, . . .	11 · 7
50 ,, ...	15 · 2		

On the 14th we sounded and dredged in 1,750 fathoms, having drawn in slightly towards the coast of Africa to get some idea of the fauna of the shallower water. The dredging was not very successful; the bottom was a dark-brown sandy ooze, with many *Globigerinæ* and other foraminifera; but beyond some fragments of a sponge, a broken sea-egg, and one or two bivalve-shells, the dredge contained no examples of the larger animal forms.

From the 15th to the 18th we continued our course, still in the Guinea Current, and under nearly similar conditions of temperature. On Tuesday, the 19th, the position of the ship at noon was lat. 5° 48′ N., long. 14° 20′ W., about two hundred miles off Cape Mesurado. A sounding was taken in the morning, in 2,500 fathoms, with a bottom of dark sandy mud. The trawl was put over, and brought up a considerable number of animal forms : among them, very prominent on account of their brilliant scarlet colour, nine large shrimps representing six species—one referred to the family of the Peneidæ, while the remainder were normal Carididæ; several tubicolous annelids, and several examples of a fine dorsibranchiate annelid with long white bristles, which, exceptionally in its class, were very distinctly jointed ; many specimens of an undescribed polyzoon with stalked avicularia and large vibracula; and a large Holothurid belonging to the gelatinous group which we had frequently met with previously in deep water, and remarkable for the position of the mouth, with its circle of branchiæ, which was placed on the lower surface of the body near the anterior extremity of the ambulatory area.

The trawl contained unfortunately only a fragment of a very large species of *Balanoglossus.*

Although evidently a worm, this animal presents so many anomalies in structure, that Gegenbaur has defined a distinct order in its class for its accommodation, under the name of the Enteropneusta. The first known species, *B. clavigerus*, was originally discovered by Delle Chiage in the Bay of Naples, and after his first description it remained long unnoticed. Kowalewski subsequently detected another species of the genus, *B. minutus*, also in the Bay of Naples; and he worked out an excellent paper on the anatomy of the genus, and showed that, like the Tunicata, *Balanoglossus* possessed a rudimentary branchial skeleton.

The body, which is worm-like, is in three marked divisions; a stout muscular proboscis, with a terminal opening for the entrance and efflux of water, round which there is a ring of rudimentary eye-spots; a strong muscular collar, somewhat like the collar in *Sabella* or *Clymene*, between which and the proboscis the mouth is placed; and the body, which is divided into three regions—first, the branchial region, which occupies about one-third of the length of the animal and in which the œsophagus is bordered by ranges of complicated gill-sacs, opening externally and supported by a delicate skeleton; secondly, a region which contains a simple stomach with hepatic cæca and the reproductive organs; and, thirdly, an enormously lengthened transparent gelatinous caudal region, terminated by the excretory opening. In our specimen only the proboscis, the collar, and the anterior portion of the branchial region were preserved; but the proportions of these— the proboscis alone eleven mm. in length by eighteen

mm. in width,—proclaimed it a giant among its
fellows.

From its structure alone *Balanoglossus* claimed a
special, we might almost say a mysterious, interest;
for its unusual branchial system—associating it, an
annelid, or perhaps more strictly an aberrant and
highly-specialised nemertid, with ascidians and with
Amphioxus—brought it into the fraternity among
which the first hazy indications of a passage between
the invertebrates and vertebrates seemed inclined to
dawn. The singular history of its development added
to the interest which had already been excited by the
peculiarities of its structure. In his series of papers
on the development and metamorphoses of the larvæ
and young of Echinoderms, Johannes Müller figured
and described what he regarded as an echinoderm
larva under the name of *Tornaria*. A couple of
years ago Metschnikoff found reason to believe that
Tornaria was the larva not of an Echinoderm, but of
Balanoglossus; and within the last year Alexander
Agassiz has confirmed Metschnikoff's view, by tracing
all the stages of its development from *Tornaria* to the
fully-formed worm.

On the 21st of August we sounded in 2,450 fathoms,
with a bottom of brownish mud evidently coloured
by the debris from some of the small rivers on the
African coast, not more than four hundred miles
distant. A temperature sounding at every hundred
fathoms down to 1,500, showed that we were still
in the Guinea Current. About mid-day we fell in
with the edge of the south-east trades, and we
shaped our course to the westward.

From the time we entered the current, immediately

after leaving the Cape Verde Islands, the sea had
been every night a perfect blaze of phosphorescence.
The weather was very fine, with a light breeze from
the south-westward. There was no moon, and
although the night was perfectly clear and the stars
shone brightly, the lustre of the heavens was fairly
eclipsed by that of the sea. The unbroken part of
the surface appeared pitch black, but wherever there
was the least ripple the whole line broke into a
brilliant crest of clear white light. Near the ship
the black interspaces predominated, but as the distance increased the glittering ridges looked closer,
until towards the horizon, as far as the eye could
reach, they seemed to run together and to melt into
one continuous sea of light. The wake of the ship
was an avenue of intense brightness. It was easy to
read the smallest print sitting at the after-port in
my cabin; and the bows shed on either side rapidly
widening wedges of radiance, so vivid as to throw
the sails and rigging into distinct lights and shadows.
The first night or two after leaving San Iago the
phosphorescence seemed to be chiefly due to a large
Pyrosoma, of which we took many specimens with the
tow-net, and which glowed in the water with a white
light like that from molten iron.

Pyrosoma is a free-swimming colony of simple
ascidians having the form of a lengthened cylinder
100 millimetres to 120 centimetres in length, with
a cavity within from 20 to 80 or 100 mm. in diameter, open at one end and closed and coming to a
point at the other; the separate individuals, often to
the number of many thousands, each included in its
proper transparent test of a consistency between

jelly and cartilage, make up the wall packed vertically side by side with all their inhalent openings
turned outwards, and the exhalent openings turned
inwards into the cavity of the cylinder. A perpetual
current is driven through each animal by the action
of the cilia bringing in freshly aërated water to a
beautifully fenestrated gill-cavity, and supplying
nourishment to a simple stomach and alimentary
tract. The consequence of this arrangement is that
the water, constantly flowing inwards through the
myriad mouths on the outer wall and finding egress
only by the open end of the cylinder, the colony is
moved steadily through the water, the closed end
first. Each animal is provided with a fairly developed
nervous system, and the whole can act in concert so
as to direct the general movements of the colony.

Besides *Pyrosoma* there were large numbers of cope-
pod crustaceans, each of which, on being shaken in the
curls of the wave, emitted a spark of light of great
intensity, and the breaking water seemed filled with
these glittering points. The tow-net brought up
during the day, but more particularly towards evening, an enormous number of pelagic animals, most of
them more or less phosphorescent. Among them,
perhaps predominating in numbers, were decapod
crustaceans in the 'zoëa' and 'megalopa' stages of
development; a great *Phyllosoma*, twelve centimetres
from tip to tip of the limbs; several species of
Leucifer; a beautiful little transparent *Cranchia*—a
cuttle-fish not more than a centimetre in length; a
Phillirrhöe, scattered over with golden spots; and an
oceanic Planarian.

As we passed southwards the character of the

phosphorescence changed somewhat. *Pyrosoma* and the larger phosphorescent creatures became less abundant, and the light given out by the water, although on the whole even more vivid than before, was more diffused, so that water shaken in a vase gave out the uniform soft light of a ground-glass globe illuminated from within by a white flame. Even when examined in small quantity in a tumbler the water was slightly turbid, and when the light was properly adjusted it was seen to contain a multitude of minute transparent bodies, which give out in the dark a clear white light becoming very vivid, almost a spark, when they are shaken or irritated.

The largest of these are spherical, nearly a millimetre in diameter. They consist of a delicate external pellicle, so thin that it can scarcely be defined under the microscope, but apparently siliceous, for when the little globe is pressed with extreme delicacy between the finger and thumb the wall of the cell is felt to break like an infinitely thin wall of glass. When the sphere is shaken from the towing-net, it usually contains only a clear transparent liquid, with a small irregularly-outlined mass of yellowish-brown sarcode sticking apparently against the inside of the cell-wall. If it be left at rest for a time in sea-water the sarcode begins to send out prolongations which gradually spread in a network of anastomosing streams over the inside of the wall, and in these streams the peculiar and extremely characteristic flowing movement of living protoplasm may be observed, each stream bearing along with it oil globules and minute granules, as in the well-known ' cyclosis ' within the cells of the moniliform hairs in the flower of *Tradescantia.* Under a high power the protoplasm is seen to consist of a clear viscid liquid, moving along with a defined edge separating it from the general fluid contents of the cell, and burdened with yellow granules and compound granular masses and minute oil globules and refractive particles; and near the centre there is always a large, well-defined nucleus of a somewhat denser material and of a greyish colour, which dyes freely with carmine. This little organism, to which Mr. Murray has given the name

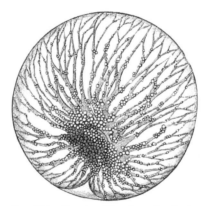

Fig. 21.—*Pyrocystis noctiluca*, Murray.   From the surface in the Guinea Current.
One hundred times the natural size.

of *Pyrocystis noctiluca* (Fig. 21), seems hitherto to have escaped notice, or if observed it has probably passed for the encysted condition of *Noctiluca miliaris,* which at first sight it greatly resembles. It certainly has nothing whatever to do with the true *Noctiluca,* which according to our observations appears to be confined to the neighbourhood of land.

Another species, *Pyrocystis fusiformis* (Fig. 22), which seems not to be quite so abundant, although

PLATE XXI. The Track of the Ship from Station 102 to San Salvador.

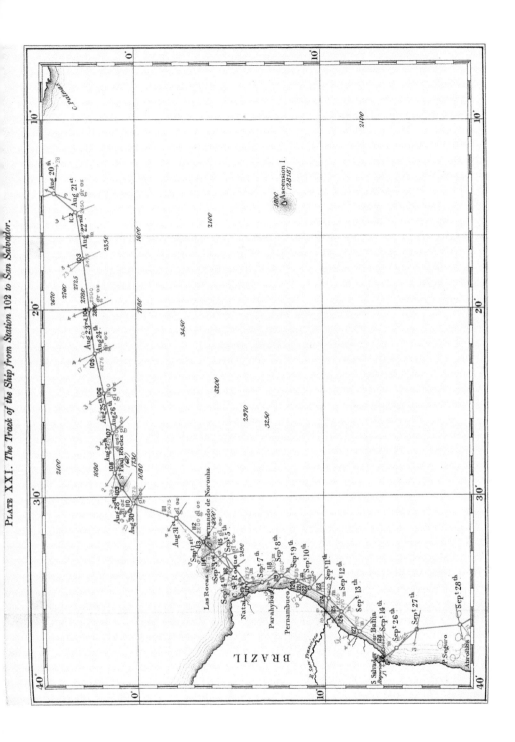

The material originally positioned here is too large for reproduction in this reissue. A PDF can be downloaded from the web address given on page iv of this book, by clicking on 'Resources Available'.

it is almost constantly associated with the preceding,
is very regularly spindle-shaped ; and a third, which
may possibly present generic differences, has the form
of a truncated cylinder. In this last we have
observed the process of endogenous multiplication

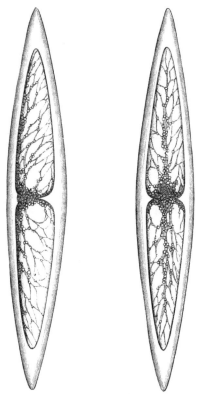

FIG. 22.—*Pyrocistis fusiformis*, MURRAY. From the surface in the Guinea Current.
One hundred times the natural size.

by the division of the protoplasmic nucleus and the
development of two secondary cells within the
parent. We are at present inclined, though with
some doubt, to relegate these forms to the Dia-
tomaceæ.

We took with the towing-net on the surface in the Guinea Current several of the *Plagusiæ,* the young flounders described by Professor Steenstrup in a remarkable paper in which he contended, though somewhat erroneously, that in passing from the young symmetrical to the adult distorted condition one of the eyes of the Pleuronectidæ passed right through the head from one side to the other. All our specimens were perfectly symmetrical, and as they ranged from one to three centimetres in length, many of them were far beyond the stage in which the wandering of the eye is described by Steenstrup, and seemed rather to favour the view that there is a group of pelagic fishes, which—while presenting all the general features of the Pleuronectidæ—never undergo that peculiar twisting which brings the two eyes of the flounder or turbot to the same side of the head, and is evidently in immediate relation with the mode of life of these animals, which feed and swim with the body closely applied to the sea-bottom.

On the 21st of August we sounded in 2,450 fathoms, with a bottom of brownish mud evidently coloured by the debris from some of the small rivers on the African coast, not more than four hundred miles distant. A temperature sounding at every 100 fathoms down to 1,500 showed that we were still in the Guinea Current. About mid-day we fell in with the edge of the south-east trades, and we shaped our course to the westward.

The depth on the 22nd was 2,475 fathoms, and the bottom temperature 1°·6 C. The position of this station was 738 miles to the eastward of St. Paul's Rocks.

The trawl was sent down on the 23rd to a depth of 2,500 fathoms, with a bottom of 'globigerina-ooze;' and during its absence temperature observations were taken at the usual intervals to 1,500 fathoms, and at every ten fathoms for the first sixty. The trawl was fairly successful, several specimens in each group, representing the sponges, the Ophiuridea, the Holothuridea, the Annelida, the Bryozoa, the Cirripedia, the macrourous Crustacea, the lamellibranchiate and gasteropod Mollusca, and the fishes having been procured; a somewhat unusually varied assemblage from so great a depth.

On the 24th we had passed the variable boundary and were in the region of the regular trades, with a steady surface-current to the north-westward of seventeen miles a day, and we found on taking a series of temperature observations down to 500 fathoms that the isotherms were again rising. The depth was 2,275 fathoms, with a bottom of 'globigerina-ooze.'

On the 25th we sounded in 1,850 fathoms, in lat. 1° 47′ N., long. 24° 26′ W., the bottom was again globigerina-ooze, and the bottom temperature 1°·8 C. A series of temperature soundings was taken at intervals of ten fathoms for the first 100, and of 100 fathoms down to 1,500. The trawl was put over, and gave us an unusually large number of interesting forms, among others many large specimens of a fine species of *Limopsis,* several brachiopods, a small *Umbellularia,* several remarkable Bryozoa, several specimens of a species of *Archaster,* some very large examples of a *Salenia* differing apparently in some respects from *S. varispina,* an entire specimen of a beautiful stalked crinoid which I shall describe under the name of *Bathycrinus aldrichianus,*[1] and with it some fragments of the stem of another form for which I propose the name *Hyocrinus bethellianus,* of which we afterwards got one or two complete specimens and several fragmentary portions, again associated with *Bathycrinus,* at Station 147, lat. 46° 16′ S., long. 48° 27′ E. about eighty-seven miles to the westward of Hog Island, one of the Crozet group. For the sake of convenience I will give a preliminary sketch of these two new crinoidal forms together.

I described and figured in 'The Depths of the Sea' (p. 452), under the name of *Bathycrinus gracilis,* a delicate little crinoid which we dredged from a depth of 2,475 fathoms to the south of Cape Clear. I believe from the structure of the stem and calyx, and from the somewhat peculiar sculpture common to both, that the first of the two forms which I have now to describe must be referred to the same genus.

In *Bathycrinus aldrichianus* (Fig. 23), the stem in full-grown specimens is 200 to 250 mm. in length, and about 2 mm. in diameter across the enlarged articulating end of one of the joints. The largest

---

[1] As the stalked Crinoids are perhaps the most remarkable of all the deep-sea groups, both on account of their extreme rarity and of the special interest of their palæontological relations, I mean to associate the names of those naval officers who have been chiefly concerned in carrying out the sounding, dredging, and trawling operations with the new species in this class, whose discovery is due to the patience and ability with which they have performed their task. Lieutenant Pelham Aldrich was first lieutenant of the 'Challenger' during the first two years of her commission; he is now with Captain Nares as first lieutenant of the 'Alert;' Lieutenant George R. Bethell, I am glad to say, was with us throughout the voyage.

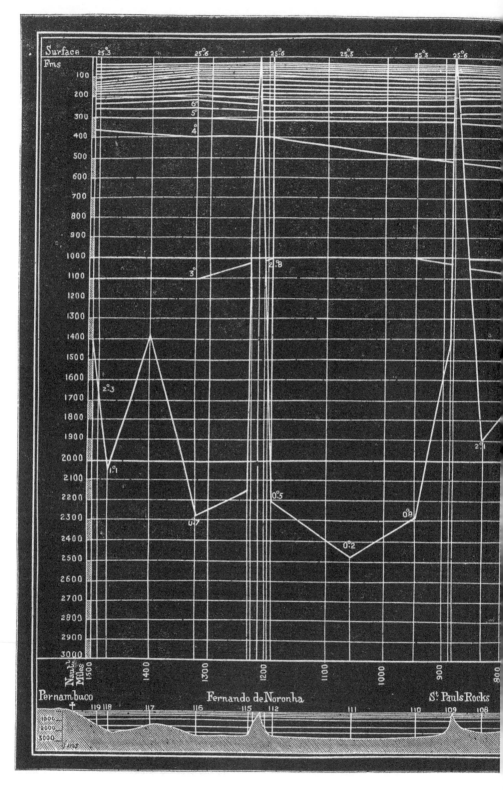

PLATE XXII.—DIAGRAM OF THE VERTICAL DISTRIBUTION OF

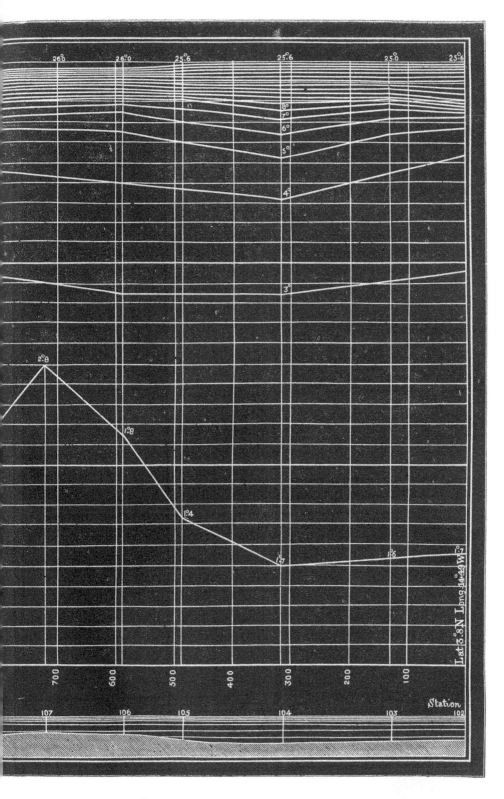

TEMPERATURE BETWEEN STATION 102 AND PERNAMBUCO.

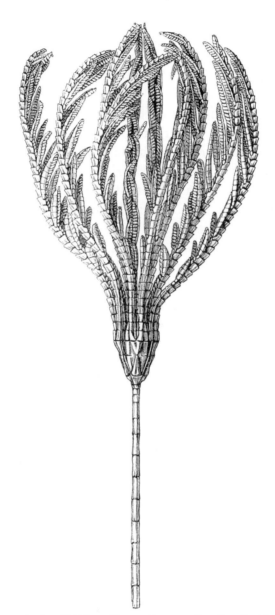

FIG. 23.— *Bathycrinus aldrichianus*, WYVILLE THOMSON. Three times the natural size.
(No. 106.)

joints of the stem have a length of about 4 mm., and they rapidly shorten towards the base of the cup. They are dice-box shaped, and have the ends bevelled off on different sides alternately, for the accommodation of masses of muscle. Towards the base of the stem a few strong jointed branches come off and form a sort of imperfect root of attachment. The cup consists of a series of basals, which are soldered together into a very small ring scarcely to be distinguished from an upper stem-joint. Alternating with these are five large triangular first radials; these are often free, but in old examples they also are frequently anchylosed into a funnel-shaped piece. The second radials are articulated to the first by a true joint with strong bands of contractile fibre; they are broad and flat, with an elevated central ridge, which is continued down upon the first radials, though in these it is not so marked; and lateral wing-like extensions, which curve up at the edges and are thus slightly hollowed out. In the third radial, or 'radial axillary,' which is united to the second radial by a syzygy, the upper border of the plate is nearly straight; but it is divided into two facets for the articulation of the two first brachials. The ridge is continued from the second radial to about the middle of the third, where it divides, and its branches pass to the insertions of the first brachials to be continued along the middle line of the arms. The wing-like lateral processes are continued along the sides of the radial axillaries, and along each side of at all events the first three brachials. The arms are ten in number; in the larger specimens they are about 30 mm. in length, and consist of from forty to fifty

joints. The first and second and the fourth and
fifth brachials are united by syzygies, and after that
syzygies occur sparingly and at irregular intervals
along the arms. There are no pinnules on the
proximal joints of the arms, but towards the distal
end there are usually about twenty, in two alternating
rows; the number and amount of development of the
pinnules seems to depend greatly on age, and not
to be very constant. The arms and the pinnules
are deeply grooved, and along the edges of the
grooves are ranges of imbricated reniform plates,
cribriform and very delicate, much resembling those in
the same position in *Rhizocrinus.* The disc is membranous, with scattered calcareous granules. The
mouth is subcentral; there are no regular oral plates,
but there seems to be a determination of the calcareous
matter to five points round the mouth, where it forms
little irregular calcareous bosses. There is an oral ring
of long fringed tentacles, and the tentacles are long and
well marked along the radial canals. The excretory
opening is on a small interradial papilla. The ovaries
are borne upon the six or eight proximal pinnules
of each arm. This form appears to be in some
respects intermediate between the pentacrinoid stage
of *Antedon* and *Rhizocrinus.* I shall reserve a discussion of its systematic position until I have an
opportunity of describing it more in detail.

Hyocrinus bethellianus (Fig. 24) is a totally different
thing; and yet from certain points of resemblance one
is inclined to regard it in the meantime as an aberrant
member of the same group. It has very much the
appearance, and in some prominent particulars it
seems to have very much the structure, of the

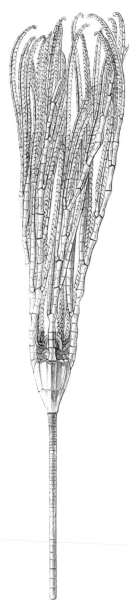

FIG. 24.—*Hyocrinus bethellianus,*
WYVILLE THOMSON. About
twice the natural size. (No. 106.)

palæozoic genus *Platycrinus* or its sub-genus *Dichocrinus.*

The longest portion of the stem which we dredged was about 170 mm. in length, but the basal part was wanting, and we had no means of determining what were its means of attachment. The stem is much more rigid than that of *Bathycrinus*, and is made up of cylindrical joints which are united to one another by a close syzygial suture, the applied surfaces being marked with a radiating pattern of grooves and ridges like those of so many of the fossil genera, and like those of the recent *Pentacrinus.* The joints become short and very numerous towards the base of the cup.

The head including the cup and the arms is about 60 mm. in length. The cup consists of two tiers of plates only. The lower of these, which must be regarded as a ring of basals, is formed, as in some of the Platycrinidæ, of two or three pieces; it is difficult to make out which with certainty, for the pieces are more or less united and the junctions in the mature animal are somewhat obscure. The second

tier consists of five radials, which are thin, broad, and
spade-shaped, with a slight blunt
ridge running up the centre and
ending in a narrow articulating surface for an almost cylindrical first
brachial. The arms are five in
number; they consist of long cylindrical joints deeply grooved and intersected by syzygial junctions. The
first three joints in each arm consist
each of two parts separated by a
syzygy; the third joint bears at its
distal end an articulating surface,
from which a pinnule springs. The
fourth arm-joint is intersected by
two syzygies, and thus consists of
three parts, and so do all the succeeding joints; and each joint gives
off a pinnule from its distal end, the
pinnules arising from either side of
the arm alternately.

The proximal pinnules are very
long, running on nearly to the end
of the arm, and the succeeding pinnules are gradually shorter, all of
them, however, running out nearly
to the end of the arm, so that distally the ends of the five arms and
of all the pinnules meet nearly on a
level. This is an arrangement hitherto
entirely unknown in recent crinoids,

Fig. 25.—*Hyocrinus be-
thellianus.* About four
times the natural size.
(Station 147.)

although we have something very close to it in some
species of the palæozoic genera *Poteriocrinus* and

Cyathocrinus. Here I believe, however, the resemblance between *Hyocrinus* and the early fossil forms ends. The outer part of the disc is paved with plates irregular in form and closely set. Round the mouth there are five very strong and definitely shaped valves, slightly cupped above, and marked beneath with impressions for the insertion of muscles. The anal opening is on a short plated interradial tube. The

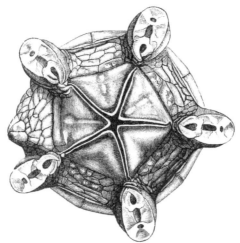

Fig. 26.—Disc of *Hyocrinus bethellianus.* Eight times the natural size.

mouth opens into a short slightly constricted œsophagus, which is succeeded by a dilatation surrounded by brown glandular ridges; the intestine is very short, and contracts rapidly to a small diameter. Round the œsophagus a somewhat ill-defined vascular ring, which may possibly be continuous with the body cavity, gives off opposite each of the oral plates a group of four tubular tentacles. The ovaries are very long, extending nearly the whole length of the first two or three pairs of pinnules on each arm.

The assemblage of characters connected with the disc and soft parts shows a considerable resemblance between *Hyocrinus* and *Rhizocrinus*. My strong impression is that the mode of nutrition of the Cyathocrinidæ, and consequently the structure and arrangement of their disc, was essentially different from that of all the yet known living forms; and I think it is probable that when we have an opportunity

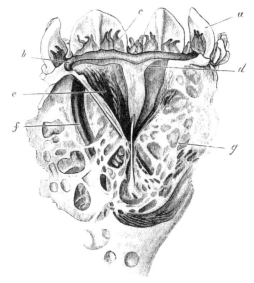

Fig. 27.—The arrangement of the soft parts in *Hyocrinus bethellianus*. *a*, oral valves; *b*, oral vascular ring; *c*, oral tentacles; *d*, *e*, inner aspect of the œsophagus and stomach; *f*, intestine; *g*, loose areolated connective tissue. Eight times the natural size.

of studying the structure of *Hyocrinus* minutely we shall find that its very striking resemblance to *Platycrinus* is to a great degree superficial.

There seems to be little doubt that *Rhizocrinus* finds its nearest known ally in the chalk *Bourguetticrinus*, and that it must be referred to the Apiocrinidæ; were it not that there is an evident relation between the two new genera and *Rhizocrinus*, in *Poteriocrinus* and *Hyocrinus* the characters of the Apiocrinidæ are so obscure that one would certainly not have been inclined to associate them with that group. They are both comparatively small forms, and although they do not show the peculiar tendency to irregularity in the number of their principal parts which we find in *Rhizocrinus*, they have still small calyces and large stems—a comparatively excessive development of the vegetative parts.

On the 27th of August we sounded in the morning in 1,900 fathoms, the bottom of little else than the shells of *Globigerina*. About two o'clock in the afternoon the look-out reported St. Paul's Rocks visible from the mast-head, and shortly afterwards they were seen from the bridge, a delicate serrated outline on the western horizon.

These solitary rocks are nearly under the equator, and midway between the coasts of Africa and of South America. They were visited by Captain Fitzroy, accompanied by Mr. Darwin, in the ' Beagle,' in 1832, and a good account of their natural history is given by Mr. Darwin in his ' Voyage of a Naturalist.' They were again touched at by Sir James Ross, in the ' Erebus,' in 1839. Merchant vessels usually give them a wide berth, but our party found a bottle with a paper, stating that on the 19th of July, 1872, Captain Pack had landed from the ship ' Ann Millicent,' of Liverpool, bound from London to Colombo. We were greatly struck with their small size, for although we knew their dimensions perfectly well—rather under a quarter of a mile from end to end of the group—we had scarcely realised so mere a speck out in mid-ocean, so far from all other land. We came in to the west of the rocks under their lee. To our right there were three small detached rocks dark and low; then a rock about sixty feet high, almost pure white, from being covered with a varnish of a mixture of phosphatic matter produced by the sea-birds and sea-salt; next a bay or cove with a background of lower rock. To the left some peaks 50 to 60 feet high, white and variously mottled, and to the extreme left detached rocks; the whole ridge excessively rugged, with channels and clefts here and there through which the surf dashes from the weather side.

A boat was sent off under the charge of Lieutenant Bethell, with a quantity of whale-line; and a loop of eight or ten ply was passed round one of the rocks. To this a hawser was run from the ship, lying about seventy yards out with her bows in 104 fathoms water. The hawser was made fast to the whale-line, and the ship thus moored to the rock. There was a strong current running past the rocks and a steady breeze blowing, both off the rocks so far as the ship was concerned, so that she was safe in any case. All was made fast about six o'clock, and Captain Nares and a small party of us went ashore in the jolly-boat. Landing on these rocks is no very easy matter. Right in the path of the trade-wind and of the equatorial current there is always a heavy surf, which had a rise and fall when we were there against the precipitous wall of rock of from five to seven feet. The rock is in rough ledges, and landing has to be accomplished by a spring and a scramble when the boat is on the top of a wave. When we landed the sun was just setting behind the ship. There was not a cloud in the sky, and the sun went down into the sea a perfect disc, throwing wonderful tints of rose-colour upon the fantastic rocks. As mentioned by Mr. Darwin, there are only two species of birds on the rocks, the 'booby' (*Sula fusca*) and the 'noddy' (*Sterna stolida*), both having a wide distribution on tropical islands and shores. On St. Paul's Rocks they are in enormous numbers, and can be seen flying round the peaks and sitting on the ledges from a great distance. We landed the first evening on the smaller rock which forms the northern portion of the ridge, and which is a breeding-place of the tern (Fig. 28). The birds were quite tame, allowing themselves to be knocked over with a stick, or even taken with the hand. They build simple nests on rocky ledges, of a conferva which grows abundantly at the water edge mixed with feathers and matted together probably with some cement matter ejected from the bird's stomach. The nests seem to be used more than once, perhaps with a little repair from time to time; for many of them were large, consisting apparently of several layers of different dates, and were decomposed at the base into a yellowish earth. A single egg was found in some of the nests, and in others a young bird, but the breeding-season was evidently nearly over. The young bird is covered with fine black down, and looks like a little ball of black wool.

The captain's party laid a line across the mouth of the cove to make landing easier for their successors, and in the evening a boat went off with officers and men to fish. The fish were in great numbers,

PLATE XXIII.—THE 'CHALLENGER' AT ST. PAUL'S ROCKS.

particularly a species of the genus *Caranx*, called, apparently in common with many other edible fishes in Spanish or Portuguese waters, 'Cavalão.' The texture of the fish is rather coarse, but the flavour is good ; it is allied to the 'tunny' of the Mediterranean.

Next day the rocks were alive with surveyors and observers of all kinds, and blue-jackets fishing and scrambling, and otherwise stretching their legs and enjoying a firm foundation under their feet. The attention of the naturalists was chiefly directed to the southern rock, which is considerably the larger. Both the tern and the booby breed here. The booby lays a single egg on the bare rock. There were a number of eggs and young birds seen, but as with the tern, the principal breeding season was past. In the morning both the booby and the noddy were quite tame, but towards afternoon even these few hours' contact with humanity had rendered them more wary, and it was now no longer possible to knock them down with sticks or stones. We had even some little difficulty in getting a specimen or two of the *Sula* for preservation, as we had unwisely left this to the last.

While some of the party were exploring the rock, we tried once more a plan of dredging which we had adopted with some success anchored on a bank at Bermudas. We sent a boat off with the dredge to a distance of a quarter of a mile or so from the ship,— the boat taking the dredge-line from the coil in the ship ; let down the dredge there and wound the dredge-rope slowly on board with the donkey-engine, thus dragging the dredge for a certain distance over the bottom. Life did not seem very abundant, but a handsome *Cidaris*, a species of *Antedon*, some crustaceans of ordinary shallow-water types, and some

Fig. 28.—Breeding-place of the Noddy. St. Paul's Rocks. (From a photograph.)

fine *Gorgoniæ* were brought up. On going over the collections from the rock, we found them to consist of a minute moth, two very small dipterous insects, a tick parasitical on the birds, a species of *Chelifer*, and three spiders. All these species had been observed previously by Mr. Darwin, with the exception of the *Chelifer*, and in addition a wood-louse and a beetle, neither of which we detected. All the insects and Arachnida were found in the old nests of the tern, many of which were brought on board and carefully examined.

There is not a trace of a land-plant on this island — not even a lichen. In the line within the wash of the surf there is a bright-pink band of an encrusting nullipore, which here and there becomes white, and greatly resembles a coral ; and the same belt produces the conferva of which the terns' nests are built, and one or two red algæ. All the crannies in the rock are inhabited by *Grapsus strigosus*, an amphibious crab, which we had already met with on several of the Atlantic islands. Its habits amused us greatly. It was much more wary than the birds—it was by no means easy to catch them, but they kept close round the luncheon baskets in large parties, raised up on the tips of their toes and with their eyes cocked up in an attitude of the keenest observation, and whenever a morsel came within their reach there was instantly a struggle for it among the foremost of them, and they ambled away with their prize wonderfully quickly with their singular sidelong gait, and a look of human smartness about them which has a kind of weirdness from its being exhibited through a set of organs totally different in aspect from those to which we usually look for manifestations of intelligence.

The lobster-pots were down during the night, but they yielded little except a small species of *Palinurus*.

The structure of the rocks is peculiar, and they must be carefully analysed before any definite opinion can be arrived at with regard to them. They are certainly, as Mr. Darwin has already pointed out, not of modern volcanic origin like almost all the other ocean islands. They look more like the serpentinous rocks of Cornwall or Ayrshire, but from these even they differ greatly in character. Mr. Buchanan examined their mineral character carefully, and subjected the most marked varieties to a rough chemical analysis. I quote from his notes. The white enamel-like encrustation described by Mr. Darwin was observed on the southern rock only, the haunt of the booby. The northern rock is chiefly composed of what appears to be Darwin's yellowish harsh stone, split up into numerous fragments which somewhat resemble large weathered crystals of orthoclase. All these rocks give off alkaline water when heated in closed tubes, and consist chiefly of hydrated oxide of magnesia, with alumina and peroxide of iron in subordinate quantity. Of the more recent veins mentioned by Darwin, some are bordered on both sides by black bands of a hard infusible substance. The powder has a dirty greyish green colour, and effervesces with dilute hydrochloric acid, leaving a brown insoluble residue. In strong hydrochloric acid it dissolves with evolution of chlorine, and the colour phenomena of dissolving peroxide of manganese. It was found to consist of phosphate of lime, peroxide of manganese, a little carbonate of lime and magnesia, and traces of copper and iron; like the other rocks, it gives off alkaline water in a closed tube. Mr. Buchanan is inclined to regard all the

rocks as referable to the serpentine group. So
peculiar, however, is the appearance which it presents,
and so completely and uniformly does the phosphatic

FIG. 29.—St. Paul's Rocks.   (From a photograph.)

crust pass into the substance of the stone, that I felt it
difficult to dismiss the idea that the whole of the crust
of rock now above water might be nothing more than

the result of the accumulation, through untold ages, of the insoluble matter of the ejecta of sea-fowl, altered by exposure to the air and sun, and to the action of salt and fresh water, but comparable with the " stalactitic or botroydal masses of impure phosphate of lime " observed by Mr. Darwin at Ascension. "The basal part of these had an earthy texture, but the extremities were smooth and glossy, and sufficiently hard to scratch common glass. These stalactites appeared to have shrunk, perhaps from the removal of some soluble matter in the act of consolidation, and hence they had an irregular form." The composition of the minerals at St. Paul's Rocks did not seem, however, to be consistent with this mode of production.

On the morning of the 29th we landed a party of explorers and fishermen, and then cast off the hawser and went round the rocks taking soundings and swinging for the errors of the compasses; and in the evening, after picking up our stragglers, we proceeded under all plain sail towards Fernando Noronha.

On the 30th we sounded in 2,275 fathoms with a bottom temperature of $+ 0°\!\cdot\!9$ C., at a distance of 265 miles to the east of Fernando Noronha, and on the 31st, at a distance of 132 miles from the island, in 2,475 fathoms, with a bottom temperature of $+ 0°\!\cdot\!2$ C. These were considerably the lowest temperatures which we had met with since the commencement of the voyage, and at first sight it seemed singular finding them almost directly on the equator. During our outward voyage circumstances prevented our tracing the source of this unusually cold water, and it was only on our return that we had an opportunity of determining that a deep indraught of cold

water passing up a channel roughly parallel with
the coast-line of South America, is open without
any intervening barrier from the southern sea to
the equator.

Early in the morning of the 1st of September, the
island of Fernando Noronha was in sight, and all
forenoon we approached it under steam, sounding at
eight A.M. in lat. 3° 33′ S., long 32° 16′ W., in
2,200 fathoms, with a bottom of globigerina-ooze,
and a bottom temperature of + 0°·5 C., the island
distant twenty-one miles. We took a series of
temperature soundings at every ten fathoms, down
to sixty fathoms—

Surface	25°· 6 C.	50 fathoms	17°· 3 C.
10 fathoms	23 · 9	60 „	15 · 0
20 „	25·· 6	75 „	13 · 6
30 „	25 · 3	100 „	12 · 4
40 „	22 · 9		

and at every hundred fathoms to 1,500. At mid-day
we sounded again about six miles from the island,
with a depth of 1,010 fathoms and a bottom temperature of 2°·8 C., so that Fernando Noronha, like most
of the ocean islands, rises abruptly from deep water.

It was a fresh, bright day, with a pleasant breeze
from the S.E. At three o'clock we cast anchor in
San Antonio Bay, just opposite the settlement
and citadel. From this point the island has a very
remarkable appearance. The land is generally not
very high, an irregular cliff rising from the sea to a
height of about a hundred feet, succeeded by undulating land and conical hills, usually covered with
luxuriant vegetation. A little to our right there was a
very singular-looking mountain, the Peak. A broad,

craggy base rises abruptly from the sea, all the clefts
among the rocks covered and filled with low vegetation, and every here and there lines and patches of
bananas. From a height of about four hundred feet a
column of rock starts up for six hundred feet more,
the last two hundred feet certainly inaccessible. On
one side there is a great cleft undercutting a projecting
portion of the rock, and adding to the grotesqueness
of its outline. The citadel, a small fort the station
of the guard of Brazilian soldiers, is on the top of a
projecting square cliff, right before the anchorage.
The village occupies a slight depression between the
citadel and the Peak, and follows the depression a good
way landwards. There is a little bit of sandy beach to
the right of the citadel, just below the village, which
is the usual landing-place ; and to the left of the
citadel (from the ship) there is a rather long stretch
of sand, with another landing, in ordinary weather
better than that near the town. To the extreme left
there is a chain of small islands, one of them with
a fine, bold outline called St. Michael's Rock, and
another much larger, flat and rather bare, Rat Island.
The view to the right is closed in by two very peculiar
conical detached rocks, called ' The Twins.'

The captain and I went ashore in the galley to pay
our respects to the governor, and to see how the land
lay. There was a heavy sea rolling on the rocks and
beaches. Some queer little catamarans were moving
about, each with a man on it, a stool, a round basket,
and a coil of fishing-line. The man either sat on the
stool or stood and propelled or guided his frail boat
with one spade-like paddle, which he plied alternately
on either side. Almost the whole of the boat, which

consists simply of two logs of wood fastened together with cross-pieces, is below the surface, and three or four of those fellows, with their scanty garments—usually reduced to a pair of short drawers—and their smooth dark skins, look oddly as if they were running about on the water without any support. One of the catamaran men spoke to us in English, and we attached him to us as interpreter, and told him to go before us to the far landing-place and then guide us to the governor's quarters. Finding the sea running so high at the landing-place as to be scarcely safe for a ship's boat, we pulled along the shore, and, taking advantage of a lull between the breakers, we ran the boat up on the far beach, and sprang out beyond reach of the surf. The road to the town lay in a hollow beyond the sea cliff. The road was tolerably good, some part of the way through sand and gravel with a tangle of bushes, most of them covered with thick masses of the long yellowish stems of the parasitical *Cuscuta americana.* Among them was growing here and there *Jatropha urens,* one of the most noxious of the island plants, stinging like a nettle, only much more bitterly. On the sides of the road the scrub became very dense, Euphorbias and leguminous plants, covered with a tangle of creepers belonging to many genera of the Circurbitaceæ, the Convolvulaceæ and Leguminosæ. The flowers of most of these were over, but still some pretty blue tufts of pea-bloom were scattered over the trees, and a little cucumber was abundantly covered with pale yellow flowers and scarlet fruit.

Near the village the road crosses a ravine, along the sides of which there are some fine banyan-trees.

A pretty little dove was in myriads in the woods. They were so tame that they would scarcely rise until we came close up to them, and if we clapped our hands they rose in a cloud, hovered in the air for a moment, and then settled down again.

On the way our guide gave us some information about the place, which we found on further inquiry to be correct. Fernando Noronha is a penal settlement, belonging to Brazil. There were then on the island the usual number of about 1,400 convicts. To hold them in check there are 200 soldiers, a governor, who holds the rank of a major in the army, and one or two other officials, with their families. Beyond these there are no inhabitants on the island, with the exception of the wives of some of the convicts, and a few women.

The usual terms of penal servitude range from five to fourteen years. The prisoners in this establishment are chiefly of a low grade, and most of them are convicted of heavy crimes,—crimes of all kinds, except, so far as we could learn, political offences. In Brazil the crime of murder is nominally punished with death, but the sentence is usually, if not always, commuted to one of transportation for life. A large number of the Fernando Noronha convicts are under these mitigated sentences. The convicts enjoy a considerable amount of liberty, and their life does not seem by any means one of great rigour. They are allowed to build a hut, and to cultivate a little piece of garden-ground on their own account, and to sell the produce. Their time and labour, from six A.M. to four P.M., belong to Government, and during that time they cultivate, in gangs on Government land,

PLATE XXIV. *Meteorological Obser*

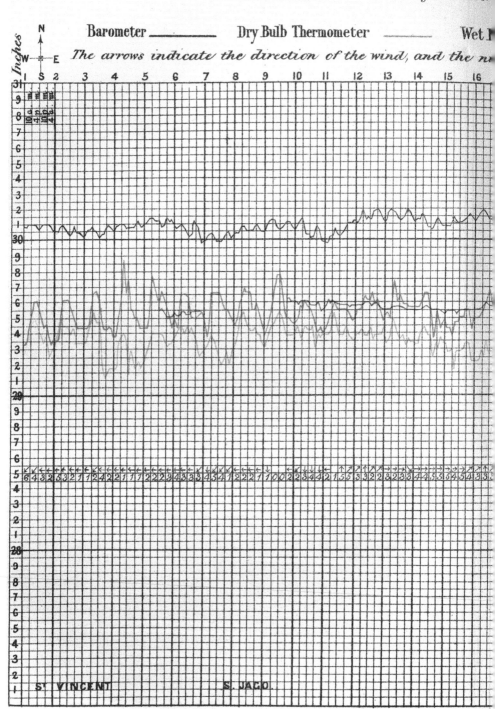

Barometer _____   Dry Bulb Thermometer _____   Wet

The arrows indicate the direction of the wind, and the n

b Thermometer _____ **Temperature of Sea Surface** _____

bers beneath its force according to Beaufort's scale

ST PAUL ROCKS.

chiefly a small black bean, on which, as it seems, they themselves chiefly subsist; and maize, which is exported about monthly to Brazil, in a little government steamer which likewise brings supplies to the island. The convicts receive from government each about 6s. a month, and have to keep themselves in food. Those who are expert fishers are allowed to ply their craft along the shore, a single man to a catamaran, and a certain proportion of the take goes to the government officials. There is no boat on the island, even in the hands of the authorities.

We were all extremely anxious to work up this island thoroughly. From its remarkable position nearly under the equator, 190 miles from the nearest land, participating, to some extent, in the conditions of the other isolated Atlantic groups, and yet, as we were well aware, in all its biological relations mainly a South American colony, it presented features of special interest to European naturalists; and it seemed to be of a size which made it possible in a few days to exhaust at all events the main features of its natural history. Accordingly we arranged parties of civilians to take up different departments, and the officers of the naval staff who were not occupied in surveying volunteered to join them and help them in collecting.

In the centre of the village in an open space with a few fine bread-fruit trees there is a solid building, forming a hollow square, which seems to be used chiefly as a prison for convicts guilty of offences on the island, and partly also as a market. Near this building a few irregular but rather neat-looking houses lodge the governor and the government officials.

We found the governor a grave, rather saturnine Brazilian, silent, partly because he spoke no foreign language and we could only communicate with him through an interpreter, and partly, I think, by nature. He asked a number of questions which surprised us a good deal from a man in his position. He inquired repeatedly what port in England we had sailed from, and to what English port we meant to return. He did not seem to understand our flag nor the captain's uniform, and asked if the ship had a commission from the British Government. He did not seem to be quite able to grasp the idea of a man-of-war for scientific purposes, and without her guns. He was very civil, however, gave us coffee and cake, and told us that we might do what we liked on the island in the way of shooting, making collections, putting up marks for surveying, &c., and offered us horses and all the aid in his power. We left him with the understanding that we were to get guides from him on the following morning and regularly to begin our work. After our interview Captain Nares and I wandered through the settlement. Irregular 'streets' or double ranges of huts radiate from the central square. The huts are all separate, each with its little garden. They are all nearly on the same plan, built of bamboo-wattles and clay, and thatched. Bananas grow wonderfully luxuriantly, embowering the little huts, some of which are whitewashed and clean and very picturesque. Often a great pumpkin plant had grown all over the roof, and loaded it with its large fruit. In the gardens there were water-melons and pump-kins, sweet potatoes, casava, lentils, and a few lemon, orange, and bread-fruit trees. The convicts were everywhere most civil; they were generally rather good-looking fellows. The great majority were of various shades of black, and often with the jolly expression so common in the different mixtures of the Negro race. In some of the huts there were women and children, and from many of them came sounds of singing and laughter, and the music of a guitar or banjo. It was difficult to realize that the whole place was a prison, with a population of convicted felons and their warders.

Beyond the village we came to some old cane-brakes, and all round there was an incessant chirping of an infinite number of crickets, not unlike our English species. They ran over the road in all directions, and one could see dozens at a time. The cane-brakes were full of doves, which rose as we approached, and fluttered up to the tops of the canes and looked at us; a little field-mouse was very abundant, scuttling about on the path and among the dry leaves; altogether, the place seemed to be very full of varied life. We walked over to the other side of the rise, and had a splendid view of the weather coast, with the curiously-formed rock, the ' clocher,' right beneath us, and the surf breaking over outlying rocks. There were some pretty views from the high ground, through cultivated valleys, dotted with banyan and bread-fruit trees and groups of palms, with scattered habitations of convicts half hidden among the beautiful foliage of the banana.

The galley had been sent off, and was to have returned for us after the men had got their supper, and one of the cutters had come on shore for the other officers. The darkness falls in these latitudes

I 2

like a curtain, and it was getting dark when we reached the beach. The captain had to look after the embarcation of the party, as the cutter was a bulky boat not well suited for surf work and had to lie out a little way. We all went off in the cutter instead of waiting for the galley, and had simply to watch for a favourable moment and make a rush for it up to the middle. We caught only one light breaker, and were soon all floundering in the boat, amid a storm of laughter.

Early next morning, when all our preparations were completed and our working-parties ready to land, Captain Nares announced that the governor had changed his mind, and did not wish to have the island examined. The captain went ashore to expostulate, and as we hoped that the change might have arisen from a misunderstanding which might be removed, boats went off with several exploring parties, the boats to lie off until one or other of two signals should be made from the ship—either the fore-royal shaken out, in which case all was to proceed as had been previously arranged; or the main-royal shaken out, when all the boats were to return to the ship. Time wore on. My *rôle* for the day was to take the steam-pinnace and dredge in moderate water off the coast. As the governor could not well object to that, I was not to be interfered with in any case, so I only waited to get a derrick fitted in lieu of one which had been damaged. About half-past ten the main-royal was shaken out, and the general recall for boats hoisted.

The pinnace had just started, and we ran back to hear the news. The governor was courteous, but

obdurate.   We might land; he would give us horses
and guides, every possible accommodation; we might
even shoot pigeons, but we must do no scientific
work.   Captain Nares asked, if we saw a butterfly,
might we not catch it, but he said he would prefer
that we should not.   The governor of a convict establishment is in a very delicate position, and bears a
heavy responsibility, not unaccompanied with serious
risk, and it is, of course, difficult to judge his conduct
in such a case; but it is not easy to see why his
determination should have been exerted against
our throwing light upon the natural history of the
island only.   Captain Nares and a party visited St.
Michael's Mount and ' Rat ' and ' Platform ' Islands.
Mr. Moseley collected a great many plants, and Mr.
Buchanan made some observations on the geological
structure of the islands, which I quote from his
notes.

"The highest island, St. Michael's Mount, forms
one of the prominent peaks which are characteristic
of the group.   It is very steep and formed entirely of
phonolite, which occurs columnar at the base and
massive towards the top; on the western side where
we landed the columns are inclined to the horizon at
an angle of about 30°.   Their transverse section looks
nearly square, the corners being, however, considerably rounded off.   The columns are for the most
part slender, and their mass is of a dirty green
colour.   In this the glassy felspar crystals are arranged with great regularity, with their broadest
faces in a plane perpendicular to the length of the
column.   The steep sides of ·the Mount are covered
with loose blocks of massive phonolite, fallen down

from above and retained in position on a very steep incline by the stems of most luxuriant creeping plants. On the weathered sides of these blocks the glassy felspar crystals, and also the crystals of hornblende, though in a less degree, project sometimes to the extent of a quarter of an inch, so much more decomposable is the crypto-crystalline matrix than the crystals occurring porphyritically in it. This rock possesses in an eminent degree the characteristic property from which it derives its name : when struck with a hammer it literally rings like a bell.

"The rock is cleft from top to bottom in two planes nearly at right angles to one another. These clefts are filled up with a hard flinty-looking substance, which appears from its structure to have been gradually deposited from water trickling down the sides. Its mass is concretionary and sometimes foliated, its colour is white to yellowish-white or brownish-yellow. It scratches glass with ease, and does not effervesce with acids. Plates of two to three millimetres in thickness are quite translucent. Heated in the forceps it does not fuse, but turns perfectly white and is then easily crumbled between the fingers, and in the closed tube it gives off alkaline reacting and empyreumatic smelling water. It was found to consist of phosphate of alumina and iron, with some silicate and sulphate of lime.

" Rat Island is the largest of the secondary islands, and the most distant from the main island. It is composed on the western side of massive basaltic rock, and on the eastern of sandstone. The sandstone probably overlies the basalt, as, in its structure, it bears the marks of having been deposited in drifts,

PLATE XXV.—FERNANDO NORONHA.

and the sand is calcareous, consisting of shell débris. On the way to and from Rat Island we had to pass along the western side of Booby Island. The wave-worn cliffs showed that the island was entirely formed of the above-mentioned calcareous sandstone; no igneous rock was visible, and, as the peculiar wind-blown stratification-marks are continued below the level of the sea, it is probable that the land here is sinking, or at all events has sunk. Platform Island consists of a mass of perfect basaltic columns rising out of the water and supporting a covering of massive basalt, on which is spread out the platform of cal-careous rock on which are the ruins of a fort, and from which the island doubtless takes its name."

In the pinnace we went along the northern shore of the main island, dredging nine times, in water from seven to twenty fathoms deep. We got sur-prisingly little, only a few crustaceans, one or two star-fishes, and a pretty little *Cidaris*. We passed some very beautiful bits of coast-scenery; a series of little sandy bays with a steep cultivated slope above them, or a dense tangle of trees absolutely imbedded in one sheet of matted climbers, separated by bold head-lands of basalt or trap-tuff. There was one particularly beautiful view when we opened ' Les Jumeaux,' and had the peak directly behind them.

Farther on, the cliffs became even more precipitous, with nests of sea-birds on all the ledges; tropic birds; a beautiful little tern, snowy white, which usually flew in pairs a foot or two apart, one following all the motions of the other, like a pair of paper butter-flies obedient to the fan of a Japanese juggler. We could see these terns flying over the land, and often alighting upon the trees. The noddy was very common, and the booby in considerable numbers. High upon the cliffs we could see the nests of the frigate-bird (*Tachypetes aquila*), and from time to time one of these splendid birds moved in slow and graceful circles over our heads. We lay for some time below the cliffs, admiring the wonderful wealth of animal and vegetable life, and returned slowly to the ship.

In the meantime, some of our party had been foraging in the town, buying up what they could from the convicts; and we were glad to see a goodly pile of water- and marsh-melons, very desirable in hot weather after a long spell at sea.

On the morning of Wednesday, the 3rd of September, we weighed anchor and left Fernando Noronha; some of us who had set our hearts upon preparing a monograph of the natural history of the isolated little island, and had made all our arrangements for the purpose, were of course greatly disappointed; but, underlying our disappointment I am inclined to think that there was a general feeling of relief on leaving a place which, with all its natural richness and beauty, is simply a prison, the melancholy habitation of irreclaimable criminals.

To show the rate at which the floor of the sea sinks in the neighbourhood of these volcanic islands: at 11.40 A.M. on the 3rd we sounded in 400 fathoms, and at 1.30 in 525 fathoms, at a distance of about six miles and a half from the island; at three o'clock we sounded in 820 fathoms with a rocky bottom, at a distance of twelve miles; and at 4.40 P.M. the depth was 2,275 fathoms with a bottom of 'globigerina-ooze.'

On the 4th we sounded in 2,150 fathoms, lat. 5° 1' S., long. 33° 50' W., about ninety miles from Cape St. Roque, and again found a comparatively low bottom temperature, + 0°·7 C. ; and on the three following days we proceeded quietly under steam, sounding from time to time in the direction of Bahia, our course lying nearly parallel with the American coast, which we could sometimes see— usually a low, uninteresting range of sandy ' dunes ' the dark line of the forest occasionally visible in the background, or the horizon broken by a delicate feathery fringe of palm-trees. On the 8th of September we sounded in 2,050 fathoms with a bottom temperature of 1°·1 C. ; and in the evening we sounded in 22 fathoms, and passed within sight of the lights of Pernambuco and Olinda.

On the morning of the 9th we were off Cape Agostinho ; we sounded in 675 fathoms in a ' globigerina-ooze ' largely mixed with river mud. The haul, as usual in such moderate depths, produced a large number of diverse invertebrates and a few very interesting fishes of deep-sea types.

Among the actinozoa, this haul yielded a very beautiful new coral (Fig. 30), which has been described by Mr. Moseley under the name of *Ceratotrochus diadema.*

The corallum is white, shallow, and saucer-shaped, with a short rudimentary pedicle and a small scar of adherence. The primary and secondary costæ are prominent and serrate ; there are six systems of septa and five cycles; the whole of the septa are exsert, the primary and secondary extremely so, projecting 10 mm. above the margin of the calicle.

The quinary septa unite with the quaternary, the quaternary with the tertiary, the tertiary with the secondary. The primary septa remain free through-out their whole course to the columella; the columella is large and oval, and composed of contorted fascicular matter. The extreme diameter of the corallum between the tips of the exsert septa is 5·75 centims, the extreme height 2·15 centims.

A young living specimen of this coral had been previously dredged at Station 78, between the islands of San Miguel and Sta. Maria, at a depth of 1,000

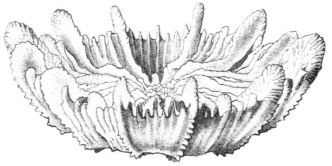

Fig. 30.—*Ceratotrochus diadema*, MOSELEY. Once and a half the natural size. (No. 120.)

fathoms; and the single adult specimen now procured was without its soft parts, but perfectly fresh and apparently only recently dead.

On the following day, keeping nearly the same course, we trawled three times at a depth of about 400 fathoms, and got a large number of very inter-esting forms, the assemblage on the whole reminding us very much of the fauna at about the same depth off the coast of Portugal. Among the special prizes were two specimens of the rare little crinoid *Rhizo-crinus lofotensis*, each infested by several individuals

of a species of *Stylifer;* and a single example of a
fine undescribed species of the genus *Pentacrinus,*
of which I shall now give a preliminary notice, pro-
posing for it the name *Pentacrinus maclearanus*[1]
(Fig. 31).

The length of the entire specimen is about 13
centims; and of these 8·5 centims are occupied by
the cup and the crown of arms and 4·5 by the stem.
As in *P. asteria* the basal joints of the stem form
interradial button-like projections, but the projecting
bosses are very evidently pointed and slightly prolonged downwards, thus showing a tendency towards
the depending processes which attain such remarkable dimensions in the liassic genus *Extracrinus.*
The first radials are low and flat—shorter in proportion to their width than in *P. asteria* and *P. mülleri;*
the second radial and the radial axillary have
much the same form and relations as they have
in the previously known species; as in *P. asteria*
there is a true joint between the first and second
radials and a syzygial junction between the second
radial and the radial axillary. The radial axillaries
support two symmetrical first brachials which are
connected with the second brachials by a syzygy.
From this point the branching of the arms is very
uniform; each of the ten primary arms gives off as a
rule two secondary arms from the inside close to the
base. To take one arm as an example of this style of
branching: the radial axillary bears two facets right
and left for two uniform first brachials, which are

[1] I dedicate this species to Captain Maclear, R.N., whose friendly
co-operation in his important executive capacity of commander of the
'Challenger,' was of the greatest importance to us.

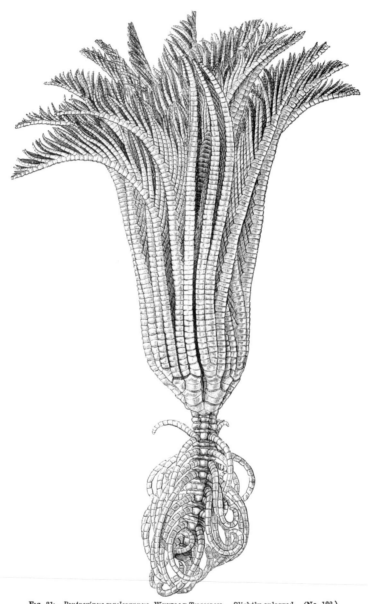

FIG. 31:—*Pentacrinus maclearanus*, WYVILLE THOMSON. Slightly enlarged. (No. 122.)

united by syzygies to brachial axillaries; these latter
have two facets of unequal size, the left facet on the
right joint and the right facet on the left joint being
small and supporting a simple arm, while the outer
facet on either joint supports a third radial, which is
connected by a syzygy with a second unequally-facetted
brachial axillary ; here again the smaller facets are
on the inside on each arm, and these give off simple
arms; simple arms spring likewise from the outer
and larger facets, but these are considerably more
robust, and are evidently the continuations of the
primary arms. Were this mode of division absolutely constant the number of arms would be thirty,
but the arrangement is slightly irregular, and in the
specimen procured thirty-one arms are present.

The arms are more regularly semi-cylindrical and
more robust than in *P. asteria*, and they have rather
a tendency to widen towards the middle of the arm.
The joints are wider and shorter than in any of the
other forms ; and the crest along the distal edge,
which is very distinct in *P. asteria*, *P. mülleri* and
P. wyville-thomsoni is scarcely perceptible. The arms
consist of about seventy joints, and there are no
true syzygies distal to the last radial axillaries.
The pinnules are comparatively broad and flat, and
consist of about fifteen joints. The disk cannot be
well seen in consequence of the attitude and rigidity
of the arms in our single example, but it appears to
resemble closely that of *P. mülleri*.

The structure of the stem is manifestly different
from that of all the hitherto described species. The
nodal joints are rather short and very much inflated,
projecting interradially in round bead-like knobs, and

the inter-nodes consist of only two very thin plate-like joints, so that the nodal joints with the rings of cirri are crowded together. The cirri start abruptly from a single nodal joint as in *P. asteria* and in *P. wyville-thomsoni;* they are robust, they consist of about twenty-five joints, and in our specimen they are closely curled downwards. From the attitude of the cirri and from the appearance of the end of the stem, there can be no doubt that this specimen is complete, that it is mature, and that it was living in an unattached condition. *Pentacrinus maclearanus* is thus very distinct from the three hitherto recognised species *P. asteria, P. mülleri* and *P. wyville-thomsoni;* perhaps it approaches the last most nearly, but it differs from it markedly in the structure and arrangement of the arms, and totally in the construction of the stem.

A 'CATAMARAN.'   FERNANDO NORONHA.

APPENDIX A.

Table of Temperatures Observed between Madeira and Station 102.

Depth in Fathoms.	Station 84. Lat. 30° 38' N. Long. 18° 5' W.	Station 85. Lat. 28° 42' N. Long. 18° 6' W.	Station 86. Lat. 25° 46' N. Long. 20° 34' W.	Station 87. Lat. 25° 49' N. Long. 20° 12' W.	Station 88. Lat. 23° 58' N. Long. 21° 18' W.	Station 89. Lat. 22° 18' N. Long. 22° 2' W.	Station 90. Lat. 20° 58' N. Long. 22° 57' W.	Station 91. Lat. 19° 4' N. Long. 24° 6' W.
Surface.	21°·7 C.	20°·7 C.	21°·7 C.	22°·5 C.	22°·2 C.	23°·0 C.	23°·3 C.	23°·3 C.
25	20·0	17·2	...	19·5	...	...	...	·'
50	17·0	16·2	...	18·0	...	...	...	...
75	16·9	15·5	...	17·5	...	...	...	...
100	16·2	14·6	...	17·2	16·1	17·7	16·7	14·1
200	13·5	12·7	...	...	13·0	13·3	13·3	12·8
300	10·9	10·7	...	...	10·2	10·2	10·0	9·8
400	9·9	9·0	...	...	8·0	7·4	7·2	7·7
500	8·7	7·6	...	...	6·8	6·8	6·8	6·2
600	8·4	...	...	...	6·0	5·7	5·4	5·3
700	6·9	...	...	...	5·2	5·2	5·2	5·0
800	5·5	...	...	...	4·7	4·7	3·1	3·3
900	4·7	...	...	...	4·2	4·1	3·6	3·2
1000	4·1	...	...	...	3·5	3·4	3·3	3·6
1100	...	...	...	...	...	2·1	3·0	...
1200	...	...	...	...	...	2·1	2·8	...
1300	...	...	...	...	...	2·6	2·4	...
1400	...	...	...	...	...	1·9	2·3	...
1500	...	...	...	...	...	2·3	...	...
Bottom Temperature.	...	...	1·8	...	1·65	1·8	1·7	1·75
Depth.	...	...	2300	...	2300	2400	2400	2075

Depth in Fathoms.	Station 92. Lat. 17° 54' N. Long. 24° 41' W.	Station 95. Lat. 13° 16' N. Long. 22° 49' W.	Station 96. Lat. 12° 15' N. Long. 22° 28' W.	Station 97. Lat. 10° 25' N. Long. 20° 30' W.	Station 98. Lat. 9° 21' N. Long. 18° 28' W.	Station 99. Lat. 7° 53' N. Long. 17° 26' W.	Station 100. Lat. 7° 1' N. Long. 15° 55' W.	Station 101. Lat. 5° 48' N. Long. 14° 20' W.	Station 102. Lat. 3° 8' N. Long. 14° 49' W.
Surface.	23° 7C.	26° 1C.	25° 9C.	25° 5	25° 7C.	25° 6C.	26° 1C.	26° 2C.	25° 6C.
25	...	...	15·7	20·6	19·2	...	...	22·6	...
50	20·0	...	12·3	15·2	15·2	17·3	16·0	17·0	18·3
75	...	...	11·8	12·7	13·0	...	...	15·0	16·1
100	16·6	11·3	10·8	11·7	12·5	13·4	12·8	13·4	13·8
200	11·1	10·1	9·3	9·4	8·7	9·8	9·7	8·8	10·4
300	8·3	7·8	7·8	6·7	7·2	7·6	6·1	6·2	5·3
400	6·5	6·0	6·2	5·7	5·3	5·5	5·4	4·8	4·7
500	5·9	4·8	5·0	4·6	5·0	5·0	4·6	4·0	3·8
600	4·8	4·5	...	4·3	...	...	4·0	3·9	4·3
700	4·4	4·8	...	4·2	...	...	3·9	3·9	3·7
800	4·5	4·2	...	3·6	...	...	3·8	3·9	3·3
900	3·6	3·6	...	3·2	...	...	3·3	...	2·4
1000	3·2	3·4	...	3·2	...	...	3·6	...	...
1100	2·8	3·1	...	2·4	...	...	2·9	...	2·9
1200	2·7	2·8	...	1·6	...	...	2·8	...	2·6
1300	1·3	2·8	...	2·2	...	...	2·6	...	2·2
1400	2·5	2·4	...	1·9	...	...	2·7	...	2·2
1500	2·7	2·3	...	2·8	...	...	3·0	...	2·2
Bottom Temperature.	...	1·8	...	1·8	2·0	...	...	1·7	1·7
Depth.	...	2300	...	2575	1750	...	...	2500	2450

Table of Temperatures observed between Station 103 and Bahia.

Depth in Fathoms.	Station 103. Lat. 2° 49' N. Long. 17° 13' W.	Station 104. Lat. 2° 25' N. Long. 20° 1' W.	Station 105. Lat. 2° 6' N. Long. 22° 53' W.	Station 106. Lat. 1° 47' N. Long. 24° 26' W.	Station 107. Lat. 1° 22' N. Long. 26° 36' W.	Station 108. Lat. 1° 10' N. Long. 28° 23' W.	Station 110. Lat. 0° 9' N. Long. 30° 18' W.	Station 111. Lat. 1° 45' S. Long. 30° 58' W.
Surface.	25°· 0 C.	25°· 6 C.	25°· 6 C.	26°· 0 C.	26°· 0 C.	25°· 9 C.	25°· 3 C.	25°· 3 C.
50	17 · 4	17 · 7	...	16 · 9	...	...	19 · 6	...
100	13 · 3	13 · 8	13 · 3	12 · 7	...	...	13 · 4	...
150	11 · 6	...	...	...	...	...	...	...
200	8 · 3	9 · 4	7 · 7	7 · 2	...	...	8 · 2	...
300	5 · 9	6 · 9	5 · 9	4 · 7	...	...	5 · 4	...
400	4 · 7	6 · 2	...	4 · 6	...	...	4 · 2	...
500	4 · 4	...	4 · 5	...	...	...	4 · 0	...
600	...	4 · 2	...	4 · 1	...	...	3 · 7	...
700	...	3 · 5	...	4 · 2	...	...	3 · 9	...
800	...	3 · 7	...	4 · 2	...	...	3 · 8	...
900	...	3 · 5	...	3 · 9	...	...	...	...
1000	...	2 · 5	. .	2 · 5	...	...	3 · 0	...
1100	...	3 · 1	...	3 · 2	...	...	3 · 2	...
1200	...	2 · 4	...	2 · 6	...	...	2 · 6	...
1300	...	2 · 4	...	...	...	...	2 · 0	...
1400	...	2 · 4	...	2 · 7	...	...	2 · 4	...
1500	...	2 · 0	...	2 · 4	...	...	2 · 2	...
Bottom Temperature.	1 · 6	1 · 7	1 · 4	1 · 8	2 · 8	2 · 1	0 · 9	0 · 2
Depth.	2475	2500	2275	1850	1500	1900	2275	2475

Depth in Fathoms.	Station 112. Lat. 3° 33' S. Long. 32° 16' W.	Station 113. Lat. 3° 50' S. Long. 32° 35' W.	Station 116. Lat. 5° 1' S. Long. 33° 50' W.	Station 118. Lat. 7° 28' S. Long. 34° 2' W.	Station 119. Lat. 7° 39' S. Long. 34° 12' W.	Station 123. Lat. 10° 9' S. Long. 35° 11' W.	Station 127. Lat. 11° 42' S. Long. 37° 3' W.
Surface.	25°· 6 C.	26°· 0 C.	25°· 6 C.	25°· 1 C.	25°· 3 C.	25°· 3 C.	25°· 0 C.
50	17 · 3	...	...	...	...	...	...
100	12 · 4	...	12 · 4	...	16 · 8	...	...
150	...	...	...	...	...	...	...
200	8 · 2	...	7 · 4	...	8 · 6	...	...
300	5 · 3	...	5 · 7	...	5 · 0	...	...
400	4 · 0	...	4 · 0	..	3 · 7	...	...
500	3 · 6	...	3 · 3	...	3 · 7	...	...
600	4 · 0	...	...	...	...	...	...
700	4 · 0	...	3 · 5	...	...	. .	...
800	4 · 2	...	3 · 5	...	...	...	...
900	3 · 4	...	3 · 5	...	...	...	...
1000	2 · 9	...	3 · 2	...	...	...	...
1100	2 · 9	...	2 · 4	...	...	...	...
1200	2 8	...	...	...	...	...	...
1300	2 · 8	...	2 · 4	...	...	...	...
1400	2 · 3	...	2 · 3	...	...	...	...
1500	2 · 5	...	2 · 8	...	...	...	...
Bottom Temperature.	0 · 5	2 · 8	0 · 7	1 · 1	2 · 3	2 · 3	3 · 3
Depth.	2200	1010	2275	2050	1650	1715	1015

APPENDIX B.

Table of Serial Temperature Soundings down to 500 Fathoms, taken between Madeira and Station 102. (Lat. 3° 8′ N., Long. 14° 49′ W.)

Depth in Fathoms.	Station 96. Lat. 12° 15′ N. Long. 22° 28′ W.	Station 98. Lat. 9° 21′ N. Long. 18° 28′ W.	Station 100. Lat. 7° 1′ N. Long. 15° 55′ W.	Station 101. Lat. 5° 48′ N. Long. 14° 20′ W.	Station 102. Lat. 3° 8′ N. Long. 14° 49′ W.
Surface.	25° · 9C.	25° · 7C.	26° · 1C.	26° · 2C.	25° · 6C.
10	21 · 4	...	26 · 1	...	25 · 8
20	17 · 1	...	25 · 6	...	23 · 9
30	...	...	23 · 0	...	23 · 0
40	...	...	17 · 9	...	20 · 8
50	12 · 3	15 · 2	16 · 0	17 · 0	18 · 3
60	...	...	...	...	15 · 8
75	11 · 8	13 · 0	...	15 · 0	16 · 1
100	10 · 8	12 · 5	12 · 8	13 · 4	13 · 8
125	10 · 7	11 · 7	...	12 · 4	...
150	10 · 0	11 · 0	...	11 · 2	12 · 6
175	9 · 8	...	...	...	...
200	9 · 3	8 · 7	9 · 7	8 · 8	10 · 4
250	8 · 6	...	...	...	...
300	7 · 8	7 · 2	6 · 1	6 · 2	5 · 3
350	6 · 7	...	...	...	...
400	6 · 2	5 · 3	5 · 4	4 · 8	4 · 7
450	6 · 0	...	...	...	...
500	5 · 0	5 · 0	4 · 6	4 · 0	3 · 8

APPENDIX C.

Table of Serial Temperature Soundings down to 200 *Fathoms, taken between Station* 102 *and Bahia:*

Depth in Fathoms.	Station 102. Lat. 3° 8' N. Long. 14° 49' W.	Station 103. Lat. 2° 49' N. Long. 17° 13' W.	Station 104. Lat. 2° 25' N. Long. 20° 1' W.	Station 106. Lat. 1° 47' N. Long. 24° 26' W.	Station 110. Lat. 0° 9' N. Long. 30° 18' W.	Station 112. Lat. 3° 33' S. Long. 32° 16' W.
Surface.	25° · 6C.	25° · 0C.	25° · 6C.	26° · 0C.	25° · 3C.	25° · 6C.
10	25 · 8	...	25 · 8	25 · 8	25 · 0	23 · 9
20	23 · 9	23 · 4	25 · 4	25 · 8	25 · 2	25 · 6
30	23 · 0	23 · 6	25 · 3	25 · 7	25 · 0	25 · 3
40	20 · 8	21 · 4	19 · 3	24 · 6	22 · 1	22 · 9
50	18 · 3	17 · 4	17 · 7	16 · 9	19 · 6	17 · 3
60	15 · 8	16 · 8	15 · 8	15 · 0	16 · 4	15 · 0
70	...	16 · 1	...	14 · 7	...	...
75	16 · 1	...	...	...	15 · 4	13 · 6
80	...	15 · 3	...	...	...	...
90	...	15 · 3	...	...	...	...
100	13 · 8	13 · 3	...	...	...	12 · 4
150	12 · 6	11 · 6	...	...	...	...
200	10 · 4	8 · 3	...	...	...	8 · 2

APPENDIX D.

Specific Gravity Observations taken between Madeira and Bahia during the months of July, August, and September, 1873.

Date 1873.	Latitude N.	Longitude W.	Depth of the Sea.	Depth (δ) at which the Water was taken.	Temperature (t) at δ.	Temperature (t') during Observation.	Specific Gravity at (t') Water at 4° = 1.	Specific Gravity at 15°.5. Water at 4° = 1.	Specific Gravity at (t.) Water at 4° = 1.
			Fms.	Fms.					
July 18	30° 38'	18° 5'		Surface.	21·7C.	22·0C.	1·02564	1·02733	1·02569
19	28 42	18 6	1125	Surface.	20·7	22·0	1·02570	1·02739	1·02603
20	27 0	19 38		Surface.	22·2	22·8	1·02563	1·02755	1·02578
21	25 46	20 34	2300	Surface.	22·5	23·2	1·02548	1·02753	1·02568
				Bottom.	1·8	20·1	1·02513	1·02631	1·02832
,, 22	23 58	21 18	2300	Surface	22·9	23·8	1·02539	1·02762	1·02562
,,				Bottom.	1·65	24·2	1·02390	1 02625	1·02827
				400	8·0	24·2	1·02396	1·02631	1·02768
,, 23	22 18	22 2	2400	Surface.	23·0	24·1	1·02494	1·02727	1·02523
24	20 58	22 57	2400	Surface.	23·3	23·9	1·02469	1·02695	1·02487
,,				100	16·7	23·5	1·02528	1·02742	1·02712
,,				150	15·0	23·5	1·02454	1·02668	1·02679
,,				300	10·0	23·5	1·02438	1·02652	1·02760
,,				400	7·2	23·5	1·02405	1·02619	1·02768
,,				500	6·75	20·7	1·02477	1·02610	1·02765
,,				Bottom.	1·8	23·5	1·02438	1·02652	1·02853
,, 25	19 4	24 6	2075	Surface.	23·6	23·9	1·02491	1·02717	1·02495
,,				100	14·1	23·9	1·02418	1·02644	1·02676
,,				200	12·8	24·0	1·02427	1·02656	1·02715
,,				300	9·8	23·9	1·02401	1·02630	1·02741
,,				400	7·7	23·9	1·02387	1·02613	1·02756
,,				Bottom.	1·75	21·0	1·02558	1·02698	1·02899
,, 26	17 54	24 41	1975	Surface.	23·9	24·6	1·02460	1·02709	1·02478
,,				45	20·0	23·4	1·02540	1·02751	1·02636
,,				75	18·5	23·0	1·02522	1·02721	1·02643
,,				100	16·6	22·8	1·02540	1·02684	1·02659
27	17 10	25 0	1070	Surface.	24·2	25·1	1·02442	1·02708	1·02470
August 6	15 43	24 15		Surface.	25·6	26·1	1·02422	1·02719	1·02434
10	13 36	22 49		Surface.	26·1	26·7	1 02374	1·02692	1·02392
			2300	Bottom.	1·8	26·1	1·02321	1·02618	1·02820
11	12 15	22 28		Surface.	25·9	26·6	1·02348	1·02663	1·02371
,,				25	15·7	25·6	1·02383	1·02664	1·02658
,,				50	12·3	25·5	1·02360	1 02640	1·02708
,,				100	10·8	25·6	1·02355	1·02636	1·02731
,,				200	9·3	25·7	1·02338	1·02622	1·02741
,,				300	7·8	25·6	1·02347	1·02628	1·02767
12	11 59	21 12		Surface.	26·1	26·5	1·02323	1·02632	1·02332
13	10 25	20 30		Surface.	25·5	26·2	1·02323	1·02623	1·02343
,,				50	15·2	26·1	1·02346	1·02643	1·02649
,,				100	11·7	25·9	1·02344	1·02633	1·02710
,,				300	6·7	26·1	1·02332	1·02629	1·02773
,,			2575	Bottom.	1·8	25·9	1·02326	1·02615	1·02816

Date 1873.	Latitude N.	Longitude W.	Depth of the Sea.	Depth (δ) at which the Water was taken.	Temperature (t) at δ.	Temperature (t') during Observation.	Specific Gravity at (t'), Water at 4° = 1.	Specific Gravity at 15°·5, Water at 4° = 1.	Specific Gravity at (t,) Water at 4° = 1.
			Fms.	Fms.					
Aug. 14	9° 15′	18° 28′		Surface.	25°·7C.	26°·5C.	1·02309	1·02618	1·02331
,,	...	...	1750	Bottom.	2·0	26·2	1·02318	1·02619	1·02816
15	7 53	17 26		Surface.	25·6	25·9	1·02322	1·02612	1·02330
16	7 1	15 55		Surface.	26·1	25·2	1·02322	1·02625	1·02326
,,	...	...		40	17·9	25·2	1·02392	1·02661	1·02599
,,	...	...		100	12·8	25·2	1·02366	1·02635	1·02690
,,	...	...		200	9·7	25·2	1·02340	1·02609	1·02721
17	6 44	16 42		Surface.	26·1	26·2	1·02337	1·02639	1·02340
18	6 11	15 57		Surface.	26·0	26·3	1·02344	1·02647	1·02350
19	5 48	14 20		Surface.	26·2	26·5	1·02336	1·02645	1·02343
20	4 29	13 52		Surface.	26·2	26·4	1·02325	1·02632	1·02330
21	3 8	14 49		Surface.	25·6	25·8	1·02314	1·02601	1·02318
,,	...	...		50	18·3	25·3	1·02398	1·02668	1·02594
,,	...	...		100	13·8	25·2	1·02385	1·02653	1·02687
,,	...	...		200	10·4	25·2	1·02362	1·02630	1·02731
,,	...	...		300	5·3	25·2	1·02352	1·02620	1·02793
,,	...	...		400	4·7	25·2	1·02373	1·02639	1·02819
,,	...	...	2450	Bottom.	1·7	25·1	1·02341	1·02606	1·02808
22	2 52	17 0	2475	Surface.	25·8	26·0	1·02338	1·02632	1·02342
23	2 25	20 1	2500	Surface.	25·6	25·8	1·02327	1·02613	1·02330
,,	...	...		Bottom.	1·7	24·9	1·02353	1·02611	1·02808
24	2 6	22 53	2275	Surface.	25·6	26·1	1·02320	1·02617	1·02333
25	1 47	24 26	1850	Surface.	26·0	26·1	1·02331	1·02628	1·02331
26	1 22	26 36	1500	Surface.	26·0	26·0	1·02332	1·02626	1·02329
,,	...	...		25	...	25·2	1·02341	1·02610	...
,,	...	...		50	...	25·1	1·02374	1·02640	...
,,	...	...		90	...	25·0	1·02375	1·02637	1·02613
,,	...	...		200	...	25·0	1·02355	1·02617	1·02797
,,	...	...		300	...	25·0	1·02366	1·02628	1·02808
,,	...	...		400	...	25·0	1·02316	1·02578	1·02759
27	1 10	28 23	1900	Surface.	25·6	25·6	1·02366	1·02651	1·02369
30	0 4	30 20		Surface.	25·3	25·8	1·02389	1·02677	1·02405
,,	...	...	2275	Bottom.	0·9	25·8	1·02327	1·02614	1·02821
31	2 6S.	31 4	2475	Surface.	25·7	26·2	1·02387	1·02692	1·02407
Sept. 1	3 42	32 21		Surface.	25·6	26·1	1·02382	1·02679	1·02396
,,	...	...	2200	Bottom.	0·5	24·3	1·02376	1·02613	1·02822
4	5 1	33 50		Surface.	25·6	25·4	1·02362	1·02635	1·02351
,,	...	...	2275	Bottom.	0·7	25·4	1·02346	1·02619	1·02827
5	4 45	33 7		Surface.	25·8	25·2	1·02402	1·02672	1·02383
6	5 54	34 39	18	Surface.	25·6	25·5	1·02404	1·02697	1·02403
7	6 38	34 33		Surface.	25·6	25·5	1·02411	1·02704	1·02421
8	7 39	34 12	1650	Surface.	25·1	25·7	1·02473	1·02760	1·02468
9	8 33	34 30	675	Surface.	25·6	25·9	1·02462	1·02752	1·02469
10	9 10	34 49		Surface.	25·3	25·8	1·02464	1·02752	1·02481
,,	...	...	400	Bottom.	...	26·3	1·02376	1·02679	...
11	10 11	35 22	1715	Bottom.	2·8	26·3	1·02378	1·02681	1·02876
12	10 46	36 8		Surface.	25·4	25·8	1·02471	1·02759	1·02484
,,	...	...	1200	Bottom.	...	26·2	1·02443	1·02749	..
13	11 52	37 10	1015	Surface.	25·0	25·0	1·02497	1·12759	1·02497

CHAPTER III.

BAHIA TO THE CAPE.

WE trawled again on the 11th in 1,715 fathoms, and this haul gave, along with a characteristic assemblage of the ordinary deep-sea invertebrates, a specimen of *Euplectella suberea*, a species which we met with first off Cape St. Vincent, and a small *Umbellularia;* and on the 12th we had two fairly successful hauls in 1,200 fathoms. Our coal was now almost entirely expended, so the engines were stopped, and on the 13th we crept along towards Bahia under all plain sail.

On the morning of the 14th of September we were steaming along the Brazilian coast towards the entrance of the magnificent Bahia de todos os Santos. All day a pretty little butterfly of the delicately-formed genus *Heliconia* was fluttering in multitudes over the ship, and over the sea as far as the eye could reach they quivered in the air like withered leaves. Their number must have been incalculable; looking up into the sky where they were thickest they were as close together and had much the appearance and style of motion of the large flakes of snow in a heavy snow-shower when a thaw is setting in. Such showers of butterflies are by no means uncommon along the coast of Brazil, nor are they confined to the Heliconidæ, although these from their extreme lightness of build seem best to fulfil the required conditions. Sometimes the country over a considerable area is absolutely devastated by some particular species of caterpillar. The butterflies or moths, as the case may be, come out nearly at one time, and the swarm of insects are caught by the land-breeze and wafted out to sea, where myriads are drowned, a remnant being, perhaps, floated back again by the usual shift of wind in the evening.

The entrance to Bahia is certainly very beautiful; we passed in the forenoon along an elevated coast, not mountainous or hilly, but rising from the shore in even terraces to the height of two or three hundred feet, the terraces broken here and there by ravines and wooded knolls, every space gloriously clothed with vegetation, and the sky-line broken by long lines of palm-trees. To the right of the town, as we

neared the anchorage, a long suburb of handsome
houses ran along the crest of the rise ; the theatre is
a prominent building in the middle of the town, and
a little above it and to the right is a handsome
church,—one with which we were afterwards very
familiar as an excellent observing station.

The general effect of the town from the sea re-
minds one somewhat of Lisbon, but Bahia is much
finer ; the splendid luxuriance of the vegetation
gives it a character of its own, and certainly nothing
approaches the palm in lightening and giving grace
to a picture.

During our stay at Bahia, Captain Maclear and
I went in one of the little coasting steamers to
Caxoeira, a small town at a few hours' distance up
a river, to get some idea of the general appearance
of the country. We were very fortunate in meeting
on board the steamer Mr. Hugh Wilson, a country-
man of our own and a leading engineer at Bahia,
who was at the time carrying out some railway
operations at Caxoeira. He had an establishment in
the town with clerks and draughtsmen at work ; there
he kindly put us up, and we rode out with him to
see the railway works. The town is on a river
between two low mountain ridges, and the railway
winds along the flank of one of these. The country
is excessively rough, with no regular roads, and it
was at first rather nervous work riding up and down
places which no civilized horses would have dreamed
of attempting. Mr. Wilson was accustomed to it
however, and led the way with the utmost confidence, and we soon learned to place complete trust
in the intelligence of the handsome black entire

horses, which seemed to be strong enough for anything and to know perfectly what they were about, often absolutely refusing to take the path indicated to them, and choosing one which to our less instructed eyes appeared ten times more difficult. In our ride we crossed here and there steep tracks winding through ravines among the mountains, and at intervals an extraordinary amount of noise, men shouting and cracking their long bullock whips, cattle struggling and scrambling among the loose boulders, and above all the shrill creaking of wheels announced the approach of one of the huge drays, dragged by ten or twelve pairs of bullocks, carrying supplies to or produce from the interior. The ponderous affair comes creaking and groaning up to the bottom of what looks like, and I suppose is, the dry bed of a torrent, and one cannot at first imagine that they can mean to attempt to go up ; after a spell of a few minutes, however, they go at it, the men shouting and lashing and every now and then putting their shoulders to the great solid spokeless wheels, and to your surprise you find that they are making a little way. One leader of a team whom we spoke to had a very confident expectation, in spite of appearances, of getting to his destination, somewhere a good way up country, in rather less than a week.

Mr. Wilson was obliged to be next day at Sto. Amaro, a little town about thirty miles distant, across one of the ridges, on another river where he had a line of steamers plying, and he asked us to ride there with him ; so we went back to his house and dined, and spent the evening at his window inhaling the

PLATE XXVI. *Meteorological Obser*

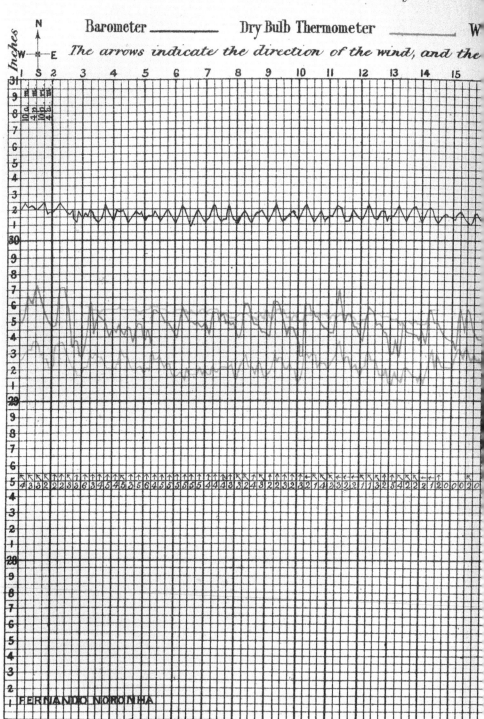

Barometer _____ Dry Bulb Thermometer _____ W

The arrows indicate the direction of the wind, and the

FERNANDO NORONHA

Bulb Thermometer ———— Temperature of Sea Surface ————

umbers beneath its force according to Beaufort's scale

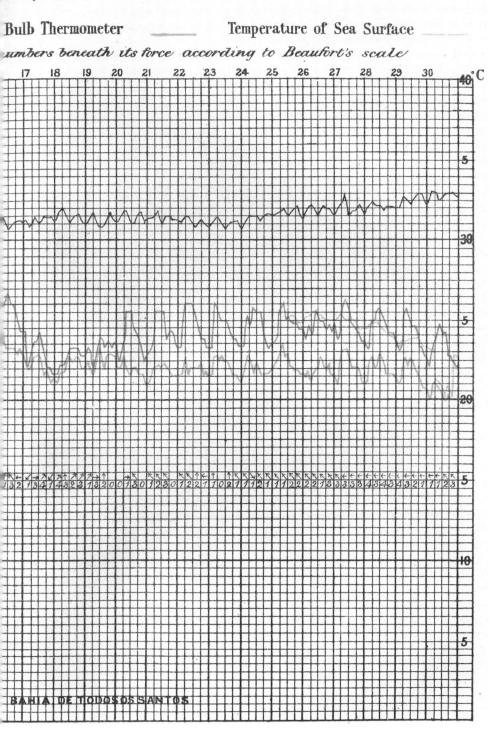

BAHIA DE TODOS OS SANTOS

soft, flower-perfumed air; and gazing at the stars twinkling in their crystal dome of the deepest blue, and their travesties in a galaxy of fire-flies glittering and dancing over the flowers in the garden beneath us. It was late when we tossed ourselves down to take a short sleep, for two o'clock was the hour fixed to be in the saddle in the morning. We rode out of the town in the star-light, Mr. Wilson, Captain Maclear, and myself, with a native guide on a fast mule. We were now obliged to trust entirely to the instinct of our horses, for if a path were visible in the day-light there was certainly none in the dark, and we scrambled for a couple of hours right up the side of the ridge. When we reached the top we came out upon flat open ground with a little cultivation, bounded in front of us by the dark line of dense forest. The night was almost absolutely silent, only now and then a peculiar shrill cry of some night-bird reached us from the woods. As we got into the skirt of the forest the morning broke, but the *réveil* in a Brazilian forest is wonderfully different from the slow creeping on of the dawn of a summer morning at home, to the music of the thrushes answering one another's full rich notes from neighbouring thorn-trees. Suddenly a yellow light spreads upwards in the east, the stars quickly fade, and the dark fringes of the forest and the tall palms show out black against the yellow sky, and almost before one has time to observe the change the sun has risen straight and fierce, and the whole landscape is bathed in the full light of day. But the morning is for yet another hour cool and fresh, and the scene is indescribably beautiful. The woods, so absolutely silent and still before, break at once into noise and movement. Flocks of toucans flutter and scream on the tops of the highest forest trees hopelessly out of shot, the ear is pierced by the strange wild screeches of a little band of macaws which fly past you like the rapped-up ghosts of the birds on some gaudy old brocade. There is no warbling, no song, only harsh noises, abrupt calls which those who haunt the forest soon learn to translate by two or three familiar words in Portuguese or English. Now and then a set of cries more varied and dissonant than usual tell us that a troop of monkeys are passing across from tree to tree among the higher branches; and lower sounds to which one's attention is called by the guide indicate to his practised ear the neighbourhood of a sloth, or some other of the few mammals which inhabit the forests of Brazil. And the insects are now all awake and add their various notes to swell the general din. A butterfly of the gorgeous genus *Morpho* comes fluttering along the path like a loosely-folded sheet of intensely blue tinsel, flashing brilliant reflections in the sun; great dark blue shining bees fly past with a loud hum; tree-bugs of a splendid metallic lustre, and in the most extraordinary harlequin colouring of scarlet and blue and yellow, cluster round a branch so thickly as to weigh it down, and make their presence perceptible yards off by their peculiar and sometimes not unpleasant odour; but how weak it is to say that that exquisite little being, whirring and fluttering in the air over that branch of *Bignonia* bells, and sucking the nectar from them with its long curved bill, has a head of ruby, and a throat of emerald, and wings of sapphire—as if any triumph of the jeweller's art could ever vie in brilliancy with that sparkling epitome of life and light!

It was broad day when we passed into the dense forest through which the greater part of the way now lay. The path which had been cut through the vegetation was just wide enough for us to ride in Indian file and with some care to prevent our horses from bruising our legs against the tree-trunks, and we could not leave the path for a single foot on either side, the scrub was so thick, what with fallen tree trunks, covered with epiphytes of all descriptions, and cycads, and arums, and great thorny spikes of *Bromelia*, and a dense undergrowth, principally of melastomads, many of them richly covered with blue and purple flowers. Above the undergrowth, the tall forest trees ran up straight and branchless for thirty or forty feet, and when they began to branch a second tier of vegetation spread over our heads, almost shutting out the sky. Great climbing *Monsteras* and other arals; and epiphytic bromeliads; and orchids, some of them distilling from their long trusses of lovely flowers a fragrance which was almost overpowering; and mazes of *Tillandsia* hanging down like tangled hanks of grey twine. Every available space between the trees was occupied by lianas twining together or running up singly, in size varying from a whip-cord to a foot in diameter. These lianas were our chief danger, for they hung down in long loops from the trees and lay upon the ground, and were apt to entangle us and catch the horses' feet as we rode on. As time wore on it got very close and hot, and the forest relapsed into silence, most of the creatures retiring for their noon-day siesta. The false roof of epiphytes and parasites kept off the glare of the sun, and it was only at intervals that a sheaf of vertical beams struck through a rift in the green canopy, and afforded us a passing glimpse of the tops of the forest trees, uniting in a delicate open tracery far above us.

For some hours our brave little horses struggled on, sometimes cantering a little where the path was pretty clear, and more usually picking their way carefully, and sometimes with all their care flounder-ing into the mud-holes, imperfectly bridged over with trunks of trees.

As we had made our ascent at first, all this time we had been riding nearly on a level on the plateau between the two river valleys. Suddenly the wood opened, and we rode up to the edge of a long irre-gular cliff bounding the valley of Sto. Amaro. The path ran right up to the edge and seemed to come to an end but for a kind of irregular crack full of loose stones which went zigzagging down to the bottom at an angle of about 70°, and we could see the path down below winding away in the dis-tance towards the main road to Sto. Amaro. We looked over this cliff and told Mr. Wilson firmly that we would *not* go down the side of that wall on horseback. He laughed, and said that the horses would take us down well enough and that he had seen it done, but that it was perhaps a little too much: so we all dismounted, and put the horses' bridles round the backs of the saddles and led them to the top of the crack and whipped them up as they do performing horses in a circus. They looked over with a little apparent uneasiness, but I suspect they had made that precarious descent before, and they soon began to pick their way cautiously down one after the other, and in a few minutes we saw them waiting for us quietly at the bottom. We then scrambled down as best we might, and it was not till we had reached the bottom, using freely all the natural advantages which the *Primates* have over the *Solidunguli* under such circumstances, that we fully appreciated the feat which our horses had performed.

The next part of the road was a trial, the horses were often up nearly to the girths in stiff clay, but we got through it somehow and reached Sto. Amaro in time to catch the regular steamer to Bahia.

At Sto. Amaro a line of tramways had lately been laid down also under the auspices of our enterprising friend, and we went down to the steamboat wharfs on one of the trucks on a kind of trial trip. The waggon went smoothly and well, but when a new system is started there is always a riks of accidents. As the truck ran quickly down the incline the swarthy young barbarians, attracted by the novelty, crowded round it, and suddenly the agonized cries of a child, followed by low moanings, rang out from under the wheels, and a jerk of the drag pulled the car up and nearly threw us out of our seats. We jumped out and looked nervously under the wheels to see what had happened, but there was no child there. The young barbarians looked at us vaguely and curiously, but not as if anything tragical had occurred, and we were just getting into the car again, feeling a little bewildered, when a great green parrot in a cage close

beside us went through no doubt another of his best
performances in the shape of a loud mocking laugh.
A wave of relief passed over the party, but we were
rather late, and the drivers expressed to the parrot
their sense of his conduct, I fear strongly, but in
terms which, being in Brazilian patois, I did not
understand.

We passed quietly down the river, with the usual
mangrove swamps and their rising background of
forest fringed with palms.   When we got outside we
found that the wind had risen, and there was a heavy
sea in the bay.   The steamer was cranky, and there
was something adrift with her engines, so we got
a good wetting before we reached the 'Challenger'
about sunset.

During our stay in Bahia the steam pinnace was
out almost daily, dredging in the shallow water, 7
to 20 fathoms in the Bay.   The fauna was wonderfully rich, every haul of the dredge bringing up
large numbers of fine tropical shore forms.   The
Echinoderms were perhaps the most striking from the
abundance of one or two large species of *Euryale* and
Antedon.   A fine calcareous sponge of unusual size
was very common; a cylindrical stem two to three
inches high supported a round button-shaped head
like an unexpanded mushroom; the regular ladder-
like arrangement of the spicules in the stem of this
species is particularly beautiful.

We remained a fortnight in Bahia and enjoyed our
stay greatly: all the conditions were so new to us
and so characteristic.   Our friend Mr. Hugh Wilson,
who was one of the leading English residents in
Bahia, and evidently a man of great energy, took us

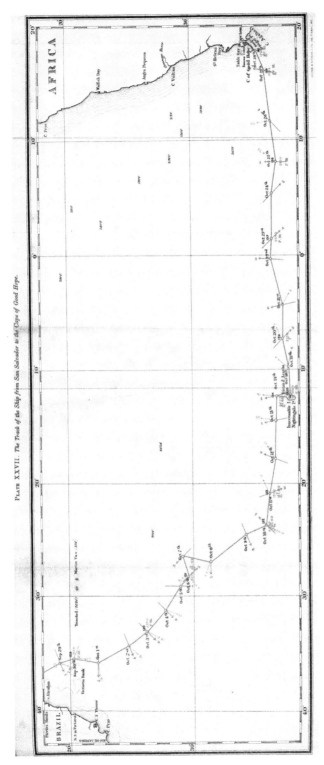

PLATE XXVII. *The Track of the Ship from San Salvador to the Cape of Good Hope.*

The material originally positioned here is too large for reproduction in this reissue. A PDF can be downloaded from the web address given on page iv of this book, by clicking on 'Resources Available'.

in charge, and very shortly an *entente cordiale* was established between our men and the young folks on shore ; and notwithstanding the broiling heat, cricketing during the day and dancing at night sped the time along.   The American frigate ' Lancaster,' arrived on the 16th, bearing the flag of Rear-Admiral Taylor, and the two crews fraternized as usual.   A ploy had been arranged for our men on board the American ship, and invitations had been issued by the ' English Cricketers ' to a ball, when we were suddenly pulled up by one of our leave-men returning on board with yellow fever.   He was at once removed to hospital on shore, but the shadow of this fell scourge having once fallen over us, no further dalliance nor delay was possible.   Leave was stopped, and as soon as the final arrangements could be made we weighed anchor and ran southwards.   The poor fellow died in hospital a few days after our departure.

Immediately outside the bay we got into fine fresh weather ; no second case appeared, and although one or two cases of simple fever which followed kept up our anxiety for a week or two, long before we reached the breezy latitudes of Tristan d'Acunha the ship was as healthy as ever, and all cause of alarm was past.

On the 26th of September we swung ship for the errors of the compasses, and for the next three days we continued our course a little to the east of south under all plain sail.   We sounded on the 30th—lat. 20° 13′ S., long. 35° 19′ W., in 2,150 fathoms, with a bottom of reddish mud, and a bottom temperature of 0°·6 C.   An attempt was made to dredge, but the dredge-rope carried away.   A serial temperature

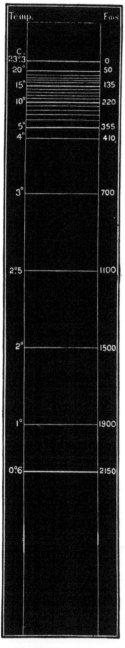

Fig. 32.—**Diagram of the Vertical Distribution of Temperature at Station 129.**

sounding was taken at intervals of 100 fathoms down to 1,500 (Fig. 32).

On the 2nd of October we saw our first albatross, sailing round the ship with that majestic careless flight which has been our admiration and wonder ever since; rising and sinking, and soaring over us in all weathers, utterly regardless of the motion of the ship, and without the slightest apparent effort. I have often watched these glorious birds for hours from the bridge, and notwithstanding all we know or think we know about the mechanics of flight, to the last I felt inclined to protest that for so heavy a bird to support itself motionless in the air, and perform its vigorous evolutions without a perceptible movement of the wings, was simply impossible by any mechanical means of which we have the least conception.

We sounded on the 3rd in 2,350 fathoms with a bottom of red mud, still due apparently in a great degree to the South American rivers, and a bottom temperature of 0°·8 C. The trawl was lowered, and on heaving in it came up apparently with a heavy weight, the accumulators being stretched to the utmost. It was a long and weary wind-in on account of the continued strain; at length it came close to the surface, and we could see the distended net through the water; when, just as it was leaving the water, and so greatly increasing its weight, the swivel between the dredge-rope and the chain gave way, and the trawl with its unknown burden quietly sank out of sight. It was a cruel disappointment,—every one was on the bridge, and curiosity was wound up to the highest pitch: some vowed that they saw resting on the beam of the vanishing trawl the white hand of the mermaiden for whom we had watched so long in vain; but I think it is more likely that the trawl had got bagged with the large sea-slugs which occur in some of these deep dredgings in large quantity, and have more than once burst the trawl net.

At 6.45 P.M. we made all plain sail, and shaped our course to the south-east.

We sounded and trawled on the 6th in 2,275 fathoms, with a muddy bottom and a bottom temperature of $0°\cdot7$ C., and obtained a series of temperature-soundings at intervals of 100 fathoms down to 1,000. The trawl came up nearly empty, containing only an ear-bone of a whale with one or two hydroid zoophytes attached to it, and a few pebbles of pumice, one having on it a large flask-shaped foraminifer or other allied rhizopod, living.

The depth on the 10th was 2,050 fathoms, the bottom an impure globigerina-ooze, and the bottom temperature $1°\cdot1$ C.: we were therefore beginning the ascent

L 2

of the western flank of the great central elevation of
the Atlantic. The temperature-determinations had
throughout the whole of this section been of the
greatest interest ; the lowest temperatures which we
had met with previously had been in the neighbourhood of Fernando Noronha, nearly under the
equator ($+ 0°·2$ C.) ; we were morally certain that
this cold water welled up from the Antarctic Sea in
the western trough of the Atlantic, and we fully
expected to intersect the line of the supply. In this
however we were disappointed ; we met with no temperature so low as the lowest temperature under the
equator ($+ 0°·2$ C.), and it was only three years
afterwards on our northward voyage that we struck
the main body of the cold indraught.

On the 11th we sounded in 1,900 fathoms with a
bottom of globigerina-ooze and a bottom temperature
of $1°·3$ C , and put over the trawl, and during its
absence took a series of shallow temperature soundings at intervals of 25 fathoms down to 100. The
trawling was comparatively successful at this station,
most of the invertebrate groups being more or less
represented. Several living specimens were procured
of a pretty little coral *Fungia symmetrica* (Fig. 33).
allied generically to the mushroom corals so abundant
in shallow water on coral reefs, and in miscellaneous
natural history collections. *Fungia symmetrica* was first
described by Count Pourtales from deep water 350 to
450 fathoms in the Strait of Florida. The corallum
is circular, plano-convex ; the wall is perfectly plane
and very little perforated, with a small convex um-
bilicus in the centre. The costæ, which correspond
with the septa, are distinct to the centre, finely

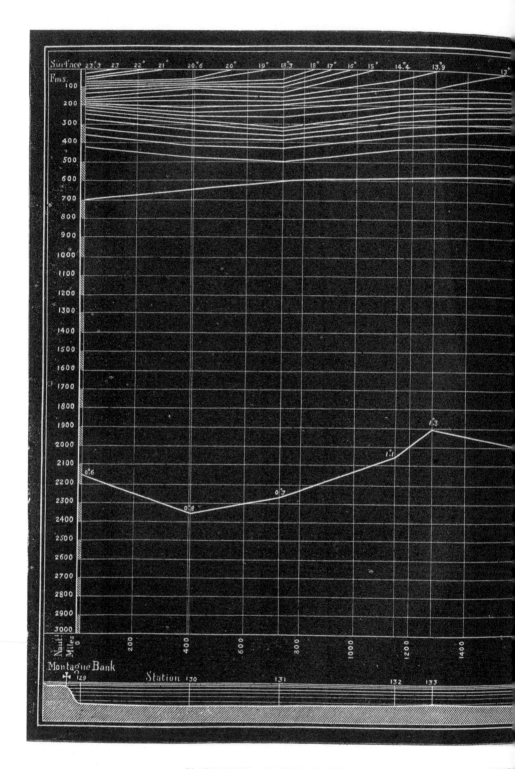

PLATE XXVIII.—DIAGRAM OF THE VERTICAL DISTRIBUTION OF TEMP

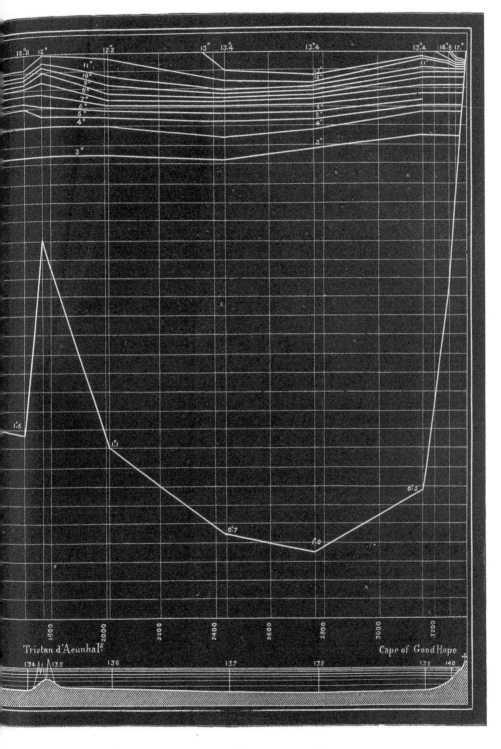

spinous and granulated, sub-equal, the primaries and
secondaries slightly the larger. The septa are sub-
equal, spinous, the larger slightly lobed; in six

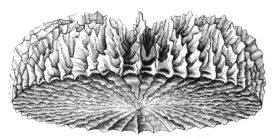

Fig. 33.—*Fungia symmetrica,* POURTALES.   Three times the natural size.   No. 133.

regular systems and four complete cycles. The septa
of the fourth cycle are connected by their inner edge
with those of the third, and the latter with those of

the second; the points of connection are sometimes expanded into a plate; the primary septa reach the centre without any connection. The columella is rudimentary, sometimes covered with a calcified membranous expansion through which some of the spines project. The synapticula are large and correspond to one another in the contiguous chambers so as to form four to six more or less regular concentric circles.

This species has been proved by our dredgings to be one of the most constantly recurring of deep-sea animals, with a world-wide distribution. It has been dredged by us fifteen times; it occurred in the North and South Atlantic, near the ice-barrier in the Southern Sea, off the West Indies, in the North and South Pacific Oceans, and among the Moluccas. It has a more extended range in depth than almost any other animal, having been obtained by us in thirty fathoms off Bermudas, and at all intermediate depths down to 2,900 fathoms. Specimens from 2,900 fathoms were obtained with the soft parts preserved, and specimens from 2,300 fathoms, of which thirty or more were obtained at one haul, were full of ripe ova. *Fungia symmetrica* is the only coral which has yet been obtained from a depth greater than 1,600 fathoms; it occurs on all kinds of bottoms—on globigerina-ooze in the Atlantic, amongst growing branched corals (*Madracis asperula*) off Bermudas, on a bottom composed almost entirely of the frustules of diatoms in the Southern Sea, and on 'red clay' with manganese nodules in the North Pacific. It sustains a range of temperature from 1° to 20° C.

The specimens from great depths are much larger

than those from shallow water, and are much more
delicate and fragile. The largest specimens procured by Count Pourtales measured one centim.
in diameter; our largest specimens were three
centim. in diameter, and those from deep water
in the North Pacific averaged two centim. The
specimens from the ' diatom-ooze' bottom, though
large, were evidently growing under circumstances
unfavourable to the formation of a corallum, the
bottom being almost entirely silicious, and only containing a trace of lime; their coralla were so fragile
that they broke with the slightest touch. From an
examination of the long series of this coral obtained
by us there seems to be no doubt of their belonging
to one species, and certain series obtained near
Bermudas and the West Indies are certainly identical
with the *Fungia symmetrica* of. Pourtales, although
some of the larger specimens seem to show close
affinities with the Lophoserinæ.

On Tuesday, the 14th of October, we sighted the
island of Tristan, distant fifty miles to the south-
south-west.

The Tristan d'Acunha group, so named from the
Portuguese navigator who discovered it early in the
sixteenth century, lies in mid-ocean, about thirteen
hundred miles south of St. Helena and fifteen hundred
west of the Cape of Good Hope, nearly on a line
between the Cape of Good Hope and Cape Horn; it
is thus probably the most isolated and remote of all
the abodes of men. The group consists of the larger
island of Tristan and two smaller islands—Inaccessible, about eighteen miles south-west from Tristan,
and Nightingale Island twenty miles south of the

main island. Tristan only is permanently inhabited, the other two are visited from time to time by sealers. We hear little of Tristan d'Acunha until near the close of last century; but even before that time it appears to have been the occasional resort of American sealers. Captain Patten, of the ship 'Industry,' from Philadelphia, arrived there in August, 1790, and remained till April, 1791. There was then abundance of wood of small growth, excellent for fire-wood, where the tents of the 'Industry's' crew were pitched, near the site of the present settlement : and the amount of sea-animals of all kinds, whales, seals, and sea-birds, was unlimited. Captain Patten's party obtained 5,600 seal-skins in the seven months of their stay, and he says that they could have loaded a ship with oil in three weeks. In 1792 the 'Lion' and the 'Hindostan,' with the British embassy to China on board, touched at the island of Tristan. The 'Lion' anchored off the north side of the island, under the cliff, but a sudden squall coming on, she almost immediately put to sea. The island was at that time entirely uninhabited ; whales and seals were seen in great numbers on the coast. In 1811 Captain Heywood found three Americans settled on Tristan preparing seal-skins and oil. Goats and pigs had been set adrift by some of the earlier visitors, and they had become very numerous on the upper terraces. One of the Americans declared himself sovereign proprietor of the islands, and in the intervals of seal-hunting they cleared about fifty acres of land, and planted it with various things, including coffee-trees and sugar-canes, which they got through the American consul at Rio. It seems that for a time some of their crops looked very promising, but for some reason the settlement was shortly abandoned. Formal possession was taken of the islands by the English in 1817, and during Napoleon's captivity at St. Helena a guard detached from the British troops at the Cape of Good Hope was maintained there. Batteries were thrown up and a few houses built, but in little more than a year the soldiers were withdrawn. A corporal of artillery of the name of Glass, with his wife and two soldiers who were induced to join him, were allowed to remain; and since that time the island of Tristan has been constantly inhabited. In 1823 the settlers were seventeen in number, among them three women, and they had to dispose of twenty-five tons of potatoes, and abundance of vegetables, milk, and butter. In 1829, when Captain Morrell visited it in the U.S. ship 'Antarctic,' the colony included twenty-seven families, and they were able to supply passing ships with bullocks, cows, sheep, and pigs, and fresh vegetables and milk in any quantity. In 1836 there was a population of forty-two on the island, and in 1852, when Captain Denham visited and sketched and roughly surveyed the group, it amounted to eighty-five, and he describes "the young men and women as partaking of the Mulatto caste, the wives of the first settlers being natives of the Cape of Good Hope and St. Helena; but the children of the second generation he would term handsome brunettes of a strikingly fine figure." They were all, at that time, members of the Church of England, under the pastoral charge of the Rev. W. F. Taylor, who had been sent out by the Society for the Propagation of the Gospel, an unknown benefactor having generously placed 1,000*l.* at the disposal of the society, to supply the colony with a clergyman for five years. Captain Denham speaks highly of the healthiness of the climate; he says that none of the ordinary epidemic diseases, whether of adults or of children, had reached the islands. The Rev. Mr. Taylor left in 1857, in H.M.S. 'Geyser,' and with him forty-seven of the inhabitants left the island and went to the Cape of Good Hope. The condition and prospects of the settlement had somewhat altered. In its early days fur-seals with pelts of good quality, inferior only to those from some of the Antarctic islands, were very abundant, and vessels could fill up at short notice with oil; it was therefore a favourite rendezvous for American sealers, and the islanders got a ready market and good prices for their produce. Gradually, however, the great sea beasts were reduced in number, the sealers and whalers had to pursue their craft further afield, and Tristan d'Acunha became only an occasional place of call. Another unfavourable change had taken place; in the early days the great majority of the population were males, but as time wore on and a new generation sprung up, the young men, scions of an adventurous stock and reared in temperance and hardihood, found their isolated life too tame for them, and sought more stirring occupation elsewhere. The proportion between the sexes rapidly altered, and at the time of Captain Denham's visit, women were considerably in the majority. The greater number of those who left Tristan in the 'Geyser' were young women, and many of them went into service at the Cape, where there still remained some of the relations of the earlier settlers.

All this time the settlement maintained an excellent character. Glass, its founder, a Scotchman born at Kelso, seems to have been a man of principle and of great energy and industry, and to have acquired to a remarkable degree the confidence of the community. He maintained his position as its leader, and represented it in all transactions with outsiders for thirty-seven years. The colony had always been English-speaking, and had strong British sympathies, and ' Governor Glass,' as he was called, had received permission from one of the naval officers visiting the island to hoist the red ensign as a signal to ships. This was the only quasi-official recognition which the colony received from Britain after the withdrawal of the troops in 1818. Glass died in 1853, at the age of sixty-seven years. He had suffered severely during his later years from cancer in the lower lip and chin, but he retained his faculties and his prestige to the last, and his death was a great loss to the little community. A general account of Tristan d'Acunha is given by the Rev. W. F Taylor, in a pamphlet published in 1850 by the Christian Knowledge Society. Mr. Taylor speaks most highly of the moral character of the flock to whom he ministered for five years; indeed, he goes so far as to say that he could find no vice to contend with, which is certainly extraordinary in so mixed an assemblage. It may be accounted for, however, to a certain degree by the compulsory sobriety of the islanders, who are usually without spirituous liquors, the rum obtained from time to time from passing ships being speedily disposed of. Mr. Taylor speaks somewhat despondingly of the prospect of the settlement. He indicates the various causes which in his opinion negative its progress, dwelling particularly upon the destruction of the wood; he looks upon the exodus which took place when he left the island as the beginning of the end, and he hopes in the interests of the settlers and of humanity that the island may soon be abandoned. Facts scarcely seem to justify Mr. Taylor's anticipations. H.R.H. the Duke of Edinburgh visited Tristan in the ' Galatea,' in 1867, and the Rev. John Milner in an entertaining narrative of the cruise gives an excellent account of the early history of the colony, and of its condition at the time of the ' Galatea's ' visit. The number of inhabitants had again risen to eighty-six, which seems to be about the normal population. Governor Glass had been dead fourteen years; he had no successor in his title; but one of the oldest of the inhabitants of the island, a man of the name of Green, who married one of Glass's daughters, had slipped into the practical part of his office, and was tacitly acknowledged as the representative of the islanders in all transactions with strangers. He lived in Glass's house, the best in the place, hoisted the red ensign and a flowing white beard, and in virtue of these symbols seemed to be accepted as general referee in all matters of difficulty. The flocks and herds were thriving, and vegetables and poultry abounded. The chaplain of the ' Galatea' christened sixteen healthy children, born since the departure of Mr. Taylor, and offered to marry seven pairs of unappropriated lads and lasses who happened oddly enough to form

PLATE XXIX.—THE ISLAND OF TRISTAN D'ACUNHA.

part of the community, but they were not inclined
to choose partners so suddenly.   The Prince and his
suite had luncheon with Mr. Green and met some of
the chief men, and all the ladies were introduced to
him.   Altogether, instead of the colony showing any
tendency to an immediate break up, there seemed to
be very general comfort and contentment.

At day-break on the 14th, the summit of the peak
of Tristan only was visible from the deck of the
' Challenger,' a symmetrical cone, the sides rising at
an angle of 23° to a height of 7,100 feet above the
level of the sea, covered with snow which came far
down occupying the ravines, dark ridges of rock
rising up between.   On account of the distance, the
lower terrace and the more level part of the island
could not be seen.   A sounding was taken in 2,025
fathoms, globigerina-ooze, the bottom temperature
1°·6 C.   The dredge was put over, and brought up
two specimens of a small *Diadema* only.   In the
evening we resumed our course towards the island,
and made all arrangements for sending out exploring
parties the first opportunity.   Early on the morning
of the 15th we were at anchor close under the land,
in a shallow bay open to the westward.   A slope of
rough pasture, about a quarter of a mile in width,
extended to our right, running up from the beach to
an almost precipitous wall of rock a thousand feet in
height, the mist lying low upon it so that we could
see no further.   To the left the rampart of rock came
sheer down almost into the sea, leaving only a narrow
strip of a few yards of shingly beach.   A stream ran
down from the high ground, nearly opposite the ship,
and the low fall with which it tumbled into the head

of the bay indicated the position of the best landing-
place. The settlement, consisting of about a dozen
thatched cottages, was scattered over the grassy

FIG. 34.—The Settlement of 'Edinburgh, Tristan d'Acunha. (*From a photograph.*)

slope, and behind it one or two ravines afforded a
difficult access to the upper terraces and the mountain. The only tree on the island is one which from

its limited distribution and the remoteness of its
locality has, so far as I am aware, no English name—
Phylica arborea. It is a small tree—allied to the
buckthorn, not rising more than twenty, or at most
thirty feet, but sending out long spreading branches
over the ground. The wood is of no value for
carpentry, but it burns well. The *Phylica* has been
exterminated on the low part of the island and in
the mouths of the ravines near the dwellings, but
there appears still to be abundance in the higher and
more distant mountain gorges. No doubt, unless
some plan is adopted for renewing the supply on the
low grounds, the labour of procuring fuel must increase, and the stock must ultimately be exhausted;
but that cannot be for a considerable time. I do not
see anything whatever in the climate, or other conditions of Tristan, to prevent the growth of the more
hardy varieties of the willow, the birch, and the
alder. The experiment is well worth trying, for the
introduction of a fast-growing hardy tree, for shelter
and for fire-wood, would increase the comfort of the
colony immensely; indeed, it seems to be all that is
necessary to insure its permanence.

A boat came alongside early in the morning, with
eight or ten of the inhabitants, some of them fine-
looking sturdy young men, somewhat of the English
type, but most of them with a dash of dark blood.
They brought a few seal-skins, some wings and
breasts of the albatross, and some sea-birds' eggs.
As it was their early spring, they had unfortunately,
with the exception of a few onions which had stood
over the winter, no fresh vegetables. Their chief
spokesman was Green, now an old man, but hale

and hearty.   He made all arrangements with the
paymaster about supplying us with fresh meat and
potatoes with intelligence and a keen eye to business.
After the departure of our guests, we landed and
spent a long day on shore, exploring the natural
history of the neighbourhood of the settlement and
learning what we could of its economy, under the
guidance of Green and some of the better informed
of the elders; while others, and more particularly
some active dark-eyed young women, got together
the various things required for the ship, each
bringing a tally to Green of her particular con-
tribution, which he valued and noted.   Most of
those who left the island in the 'Geyser' and the
'Galatea' have returned, and the colony at present
consists of eighty-four souls in fifteen families, the
females being slightly in the majority.   Most of
the settlers are in some way connected with the
Cape of Good Hope; some are Americans.   The
greater number of the women are Mulattoes.   Many
of the men are engaged in the seal and whale
fishery, and as that has now nearly come to an
end on their own shores, they are generally employed
on board American whalers in the southern seas.
We had a good deal of conversation with a son of
Governor Glass, a very intelligent handsome young
man, who had been at Kerguelen-land, and at several
other whaling stations in the south, and who gave
us some useful information.   The chief traffic of
the islanders is with these American ships, from
eight to twelve of which call in passing yearly, to
barter manufactured goods and household stuffs
for fresh vegetables and potatoes.

The fifteen families possess from five to six hundred head of cattle, and about an equal number of sheep, with pigs and poultry in large numbers. Beef was sold to our messman at 4*d.* a pound, mutton at 4*d.*, pork somewhat cheaper, and geese at 5*s.* each, so that the Tristaners, so long as they can command a market—and the number of their occasional visitors is increasing with increasing communication and commerce—cannot be considered in any way ill off. Their isolation and their respectability, maintained certainly with great resolution and under trying circumstances, induce a perhaps somewhat unreasonable sympathy for them, which they by no means discourage and which usually manifests itself in substantial gifts.

The cottages are solid and comfortable. They usually consist of two or three rooms, and are built of a dark brown tufaceous stone, which they blast in large blocks from the rocks above, and shape with great accuracy with axes. Many of the blocks are upwards of a ton in weight, and they are cut so as to lock into one another in a double row in the thickness of the wall, with smaller pieces equally carefully fitted between them. There is no lime on the island, so that the blocks are fitted on the cyclopean plan, without cement. With all precautions, however, the wind sometimes blows from the south-west with such fury that even these massive dwellings are blown down; and we were assured that the rough blocks, brought from the mountain and laid on the ground to be fashioned, are sometimes tumbled about by the force of the wind.

They have on the island a few strong spars, mostly the masts of wrecked vessels, and to get the great blocks up to the top of the wall after it has risen to a certain height, they use a long incline, made of a couple of these spars, well greased, up which they slowly drag and shove the blocks, much as they are represented as doing in old times in some of the Egyptian hieroglyphs. The furniture of the rooms is scanty, owing to the difficulty of procuring wood ; but passing ships seem to furnish

Fig. 35.—Cyclopean architecture, Tristan Island. (*From a photograph.*)

enough of woven fabrics to supply bedding, and in the better cottages some little drapery, and to enable the people, and particularly the women, to dress in a comfortable and seemly style. Low stone walls partition the land round the cottages into small enclosures, which are cultivated as gardens, and where all the ordinary European vegetables thrive fairly. There is no fruit of any kind on the island. The largest cultivated tract is on the flat, about half a mile from 'Edinburgh.' There the greater part of the potatoes are grown, and the cattle and sheep have their head-quarters. The goods of the colonists are in no sense in common.; each has his own property in land and in stock. A new-comer receives a grant of a certain extent of land, and he gets some grazing rights, and the rest of the settlers assist him in fencing his patch, and in working it and preparing it for a first crop. They then contribute the necessary cattle, sheep, potato-seed, &c., to start him ; contributions which he no doubt repays when he is in a position to do so, under some definite understanding, for the Tristan Islanders have a very practical knowledge of the value of things. There seems to be a harmonious arrangement among them for assisting one another in their work, such assistance being repaid either in kind or in produce or money. The community is under no regular system of laws, everything appears to go by a kind of general understanding. When difficulties occur they are referred to Green, and perhaps to others, and are settled by the general sense. This system is probably another great source of the apparently exceptional morality of the place ; in so small a community where all are so entirely interdependent, no misconduct affecting the interests of others can be tolerated or easily concealed, and as there is no special machinery for the detection and punishment of offences, the final remedy lies in the hands of the men themselves, who are most of them young and stalwart and well able to keep unruliness in check.

The island of Tristan is almost circular, about seven miles in diameter. The position of Herald

Point, close to the settlement, is lat. 37° 2′ 45″ S., long. 12° 18′ 30″ W., so that it nearly corresponds in latitude with the Açores and the southern point of Spain in the northern hemisphere. The island is entirely volcanic, the cliff—upwards of a thousand feet high—which encircles it, breached here and there by steep ravines, is formed of thin beds of tuffs and ashes, some of them curiously brecciated with angular fragments of basalt; and layers of lava intersected by numerous dykes of varying widths of a close-grained grey dolerite. The cone is very symmetrical, almost as much so as the Peak of Teneriffe, and the flows of lava down its flanks appear as rugged black ridges through the snow. The inhabitants sometimes go to the top, and they represent the mountain as a cone of ashes, with a lake on the summit. The upper terrace is covered with long, coarse grass, with a tangled brush of *Phylica* in the shelter of the ravines.

Two species of albatross breed on the higher parts of the island, *Diomedea exulans* and *D. chlororhynchus*, the former even beyond the summer limit of the snow. A few years ago there were large flocks of goats on the upper terraces, but latterly, from some unknown cause, they have entirely disappeared, and not even the remains of one of them can be found. With the exception of the goat and the pig, and the rat and the mouse, which are known to have been recently introduced, there are no land quadrupeds at large on the island, and the land birds, so far as we know, are confined to three species—a thrush, *Nesocichla eremita*, a bunting referred by Captain Carmichael to *Emberiza brasiliensis*, and a singular bird called by the settlers the 'island hen,' which was at one time very common, but which is now almost extinct. This is a water-hen, *Gallinula nesiotis* (Sclater, Proc. Zool. Soc. 1861), very nearly allied to our common English moor-hen (*Gallinula chloropus*), which it resembles closely in general appearance and colouring, with, however, several satisfactory specific differences. The wings of the Tristan species are much shorter, and the primary feathers, and indeed all the feathers of the wing, are so short and soft as to be useless for the purposes of flight. The breast-bone is short and weak, and the crest low, while, on the other hand, the pelvis and the bones of the lower extremity are large and powerful, and the muscles attached to them strong and full. The island hen runs with great rapidity; it is an inquisitive creature, and comes out of its cover in the long grass when it hears a noise. It is excellent eating, a good quality which has led to its extermination. Mr. Moseley collected between twenty and thirty plants on Tristan, perhaps the most interesting a geranium (*Pelargonium australe*, Var.), a species which extends, in several varieties, to the Cape, New Zealand, and Australia.

We heard a curious story at Tristan about two Germans who had settled nearly two years before on Inaccessible Island. Once a year, about the month of December, the Tristan men go to the two outlying islands to pick up the few seals which are still to be found. On two of these occasions they had seen the Germans, and within a few months smoke had risen from the island, which they attributed to their having fired some of the brush; but as they had seen

Fig. 36.—Water-Fall, Inacessible Island.　(*From a photograph.*)

or heard nothing of them since, they thought the probability was that they had perished. Captain Nares wished to visit the other islands, and to ascertain the fate of the two men was an additional object in doing so.

Next morning we were close under Inaccessible Island, the second in size of the little group of three. The ship was surrounded by multitudes of penguins, and as few of us had any previous personal acquaintance with this eccentric form of life, we followed their movements with great interest. The penguin as a rule swims under water, rising now and then and resting on the surface, like one of the ordinary water-birds, but more frequently with its body entirely covered, and only lifting its head from time to time to breathe.

One peculiarity surprised us greatly, for although we were tolerably familiar with the literature of the family, we had never seen it described. The 'rock-hopper,' and I am inclined to think species of other genera besides *Eudyptes*, when in a number in the water have a constant habit of closing together the legs and tail straight out, laying the wings flat to the sides, arching forward the neck, and, apparently by an action of the muscles of the back, springing forwards clear out of the water, showing a steel-grey back and a silvery belly like a grilse. They run in this way in lines like a school of porpoises, seemingly in play, and when they are thus disporting themselves it is really very difficult to believe that one is not watching a shoal of fish pursued by enemies.

In the water penguins are usually silent, but now

and then one raises its head and emits a curious, prolonged croak, startlingly like one of the deeper tones of the human voice. One rarely observes it in the daylight and in the midst of other noises, but at night it is weird enough, and the lonely officer of the middle watch, whose thoughts may have wandered for the moment from the imminent iceberg back to some more genial memory, is often pulled up with a start by that gruff 'whaat' alongside in the darkness, close below the bridge.

The structure of this island is very much the same as that of Tristan, only that the pre-eminent feature of the latter, the snowy cone, is wanting. A wall of volcanic rocks, about the same height as the cliff at Tristan, and which one is inclined to believe to have been at one time continuous with it, entirely surrounds Inaccessible Island, falling for the most part sheer into the sea, and it seems that it slopes sufficiently to allow a tolerably easy ascent to the plateau on the top at one point only.

There is a shallow bay in which the ship anchored in fifteen fathoms on the east side of the island; and there, as in Tristan, a narrow belt of low ground extending for about a mile along the shore is interposed between the cliff and the sea. A pretty water-fall tossed itself down about the middle of the bay over the cliff from the plateau above. A little way down it was nearly lost in spray, like the Staubbach, and collected itself again into a rivulet, where it regained the rock at a lower level. A hut built of stones and clay, and roofed with spars and thatch, lay in a little hollow near the waterfall, and the two Germans in excellent health and spirits, but enraptured at the sight of the ship and longing for a passage anywhere out of the island, were down on the beach waiting for the first boat. Their story is a curious one, and as Captain Nares agreed to take them to the Cape, we had ample time to get an account of their adventures, and to supplement from their experience such crude notions of the nature of the place as we could gather during our short stay.

Frederick and Gustav Stoltenhoff are sons of a dyer in Aix-la-Chapelle. Frederick, the elder, was employed in a merchant's office in Aix-la-Chapelle at the time of the outbreak of the Franco-Prussian war. He was called on to serve in the German army, where he attained the rank of second lieutenant, and took part in the siege of Metz and Thionville. At the end of the campaign he was discharged, and returned home to find his old situation filled up.

In the meantime, his younger brother, Gustav, who was a sailor and had already made several trips, joined on the 1st of August, 1870, at Greenock, as an ordinary seaman, the English ship ' Beacon Light,' bound for Rangoon. On the way out the cargo which consisted of coal caught fire when they were from six to seven hundred miles north-west of Tristan d'Acunha, and for three days all hands were doing their utmost to extinguish the fire. On the third day the hatches, which had been battened down to exclude the air, blew up, the main hatch carrying overboard the second mate who had been standing on it at the time of the explosion. The boats had been provisioned beforehand, ready to leave the ship.

Two of the crew were drowned through one of the
boats being swamped, and the survivors, to the number of sixteen, were stowed in the long-boat.   Up to
this time the ship had been nearing Tristan with a
fair wind at the rate of six knots an hour, so that
they had now only about three hundred miles to go.
They abandoned the ship on Friday : on Sunday
afternoon they sighted Tristan, and on the following
day a boat came off to their assistance and towed
them ashore.

The shipwrecked crew remained for eighteen days
at Tristan d'Acunha, during which time they were
treated with all kindness and hospitality.   They were
relieved by the ill-fated ' Northfleet,' bound for Aden
with coal, and Gustav Stoltenhoff found his way back
to Aix.

During his stay at Tristan he heard that large
numbers of seals were to be had among the islands,
and he seems to have been greatly taken with the
Tristaners, and to have formed a project of returning
there.   When he got home his brother had just got
back from the war and was unemployed, and he
infected him with his notion, and the two agreed to
join in a venture to Tristan to see what they could
make by seal-hunting and barter.

They accordingly sailed for St. Helena in August,
1871, and on the 6th of November left St. Helena
for Tristan in an American whaler bound on a cruise
in the South Atlantic.   The captain of the whaler,
who had been often at Tristan d'Acunha, had some
doubt of the reception which the young men would
get if they went as permanent settlers, and he spoke
so strongly of the advantages of Inaccessible Island,

on account of the greater productiveness of the soil, and of its being the centre of the seal fishing, that they changed their plans and were landed on the west side of Inaccessible on the 27th of November, —early in summer. A quarter of an hour after, the whaler departed, leaving them the only inhabitants of one of the most remote spots on the face of the earth. They do not seem, however, to have been in the least depressed by their isolation.

The same day the younger brother clambered up to the plateau with the help of the tussock-grass, in search of goats or pigs, and remained there all night, and on the following day the two set to work to build themselves a hut for shelter. They had reached the end of their voyage by no means unprovided, and the inventory of their belongings is curious.

They had an old whale-boat which they had bought at St. Helena, with mast, sails, and oars; three spars for a roof, a door, and a glazed window; a wheelbarrow, two spades and a shovel, two pickaxes, a saw, a hammer, two chisels, two or three gimlets, and some nails; a kettle, a frying-pan, two saucepans, and knives and forks, and some crockery; two blankets each, and empty covers which they afterwards filled with sea-birds' down. They had a lamp and a bottle of oil, and six dozen boxes of Bryant and May's matches.

For internal use they had two hundred pounds of flour, two hundred pounds of rice, one hundred pounds of biscuit, twenty pounds of coffee, ten pounds of tea, thirty pounds of sugar, three pounds of table-salt, a little pepper, eight pounds of tobacco,

five bottles of hollands, six bottles of Cape wine, six
bottles of vinegar, and some Epsom salts. A barrel
of coarse salt was provided for curing seal-skins, and
fourteen empty casks for oil. Their arms and ammu-
nition consisted of a short Enfield rifle, an old
German fowling-piece, two and a half pounds of
powder, two hundred bullets, and four sheath-knives.
The captain of the whaler gave them some seed-
potatoes, and they had a collection of the ordinary
garden seeds.

When they had been four days on the island they
had a visit from a party of men from Tristan, who
had come on their annual sealing excursion. They
were ten days on Inaccessible, and were very friendly
in their intercourse with the new-comers. They told
them that the north side of the island was better
suited for a settlement, and transported all their
goods thither in one of their boats. Being familiar
with the place, they showed them generally their
way about, and the different passes by which the
plateau might be reached from beneath, and they
taught them how to build to withstand the violent
winds, and how to thatch with tussock-grass.

Immediately after they left, the brothers set about
building a house and clearing some ground for
potatoes and other vegetables. They killed nineteen
fur-seals, and prepared the skins, but they were unable
to make any quantity of oil. Towards the end of the
sealing season their boat got damaged in the surf, and
they were obliged to cut it in two and patch up the
best half of it, and use it as best they might in smooth
weather close to the shore.

They went from time to time to the upper

plateau and shot goats and pigs. When they first arrived they counted a flock of twenty-three goats; three of these were killed during the summer of 1871-2 by the Tristan people, and six by themselves; the remaining fourteen remained over the winter of 1872. The flesh of the goats they found extremely delicate. Pigs were much more numerous, but their flesh was not so palatable, from their feeding principally on sea-birds; that of the boars was especially rank. They found the pigs very valuable, however, in yielding an abundant supply of lard, which they used to fry their potatoes.

In the month of April 1872 a singular misfortune befell them. While burning some of the brush below to make a clearing, the tussock in the gully by which they had been in the habit of ascending the cliff caught fire, and as it had been only by its assistance that they had been able to scramble up the plateau, their only hunting-ground was now inaccessible from the strip of beach on which their hut and garden stood, which was closed in at either end by a headland jutting into the sea. While their half-boat remained seaworthy they were able to paddle round in fine weather, to the west side of the island, where there was an access to the top; but the ' sea-cart,' as they called it, was washed off the beach and broken up in June, and after that the only way they had of reaching the plateau was by swimming round the headland —a risky feat, even in the finest weather, in these wild regions.

In winter it was found to be impossible to reach the terrace, and as their supply of food was low, they experienced considerable privations during their first

winter. Their daily allowance of food was reduced
to a quantity just sufficient to maintain life, and in
August they ' were little better than skeletons.' Help
was, however, near. Early in August a multitude of
penguins landed at a ' rookery ' hard by their hut,—
stupid animals, which will scarcely get out of one's
way, and are easily knocked down with a stick, and
with fleshy breasts, wholesome enough, if with a
rather fishy taste; and in the end of August the
females began to lay large blue eggs, sufficiently
delicate in flavour.

A French barque hove-to off the beach in the middle
of September, and in her they shipped their seal-skins,
and bartered penguins' eggs with her for biscuits and
tobacco. Had the barque arrived a week earlier the
brothers would have left the island; but the eggs had
set them up again, and they determined to remain a
little longer. In October a fore-and-aft schooner,
which proved to be the ' Themis,' a whaler from the
Cape of Good Hope, was seen standing towards the
island. A gale of wind blew her off for a couple of
days, but she returned and communicated, landing
some men from Tristan, who had crossed to see what
the hermits were about. Their guests remained a day
and a half, and returned to Tristan.

Early in November, that is, early in their second
summer, the brothers swam round the eastern headland—Frederick with their blankets, the rifle, and a
spare suit of clothes—Gustav with powder, matches,
and the kettle in an oil-cask. They mounted by the
help of the tussock-grass to the top of the cliff,
went over to the west side of the plateau, and built
a small hut, where they remained a month, living

on goats' flesh and fresh pork. On the 10th of
December they returned home, mended their thatch,
dug the early potatoes and put the garden in order.

On the 19th of December the Tristan men made
their second sealing expedition. They remained nine
days on the island, and killed forty seals, one sea-
elephant, and eight of the remaining twelve goats.
They left some flour in exchange for an oil-cask, and
this was the last communication between the brothers
and the outer world until the 'Challenger' called
eight months later. In January, Frederick swam
round the point again, and mounted the cliff. He
shot four pigs, ran the fat into buckets, and threw
the hams down to his brother on the beach below.
He saw the four last goats, but spared them to
increase their number. In February a boat came
to the west side from Tristan, and its crew killed the
four goats, and departed without communicating with
the Stoltenhoffs.

The relations between the Tristan people and the
brothers does not appear to have been so cordial
latterly as it was at first, and the Stoltenhoffs believe
that the object of their neighbours in killing the
goats, and in delaying from time to time bringing
them some live stock which they had promised them,
was to force them to leave the island. It may have
been so, for the Tristan men had been in the habit
of making a yearly sealing expedition to Inaccessible,
and no doubt the presence of the energetic strangers
lessened their chance of success.

In March the brothers once more swam round the
point and ascended the cliff. After staying on the
plateau together for a few days, it was settled that

Frederick should remain above and lay in a stock of
lard for the winter, Gustav returning to the hut and
storing it.   When a pig was killed, the hide, with the
fat in it, was rolled up, secured with thongs of skin,
and thrown over the cliff, and Gustav then ran the
lard into a cask.

During their second winter the privations of the
brothers do not seem to have been great.   They were
getting accustomed to their mode of life, and they
had always sufficient food, such as it was.   They were
remarkably well educated : both could speak and read
English fluently, and the elder had a good knowledge of French.   Their library consisted of eight
volumes : Schœdler's Natural History, a German
Atlas, Charles O'Malley, Captain Morrell's Voyages,
two old volumes of a monthly magazine, Hamlet
and Coriolanus with French notes, and Schiller's
Poems.   These they unfortunately came to know
almost by heart, but they had considerable resources
in themselves, in the intelligent interest which they
took in the ever-changing appearances of nature.

When the 'Challenger' arrived they were preparing
for another summer, but the peculiar food and the
want of variety in it was beginning to tell upon
them, for all their original stores were exhausted,
with the exception of the Epsom salts which was
untouched, neither of them having had an hour's
illness during their sojourn ; and they were heartily
glad of the chance of a passage to the Cape.

Frederick came to the ship to see us before we left
for the south in December.   He was then comfortably
settled in a situation in a merchant's office in Cape
Town, and Gustav was on his way home to see his

PLATE XXX. *Meteorological Obser*

Barometer ——————— Dry Bulb Thermometer ——————— Wet

The arrows indicate the direction of the wind, and the n

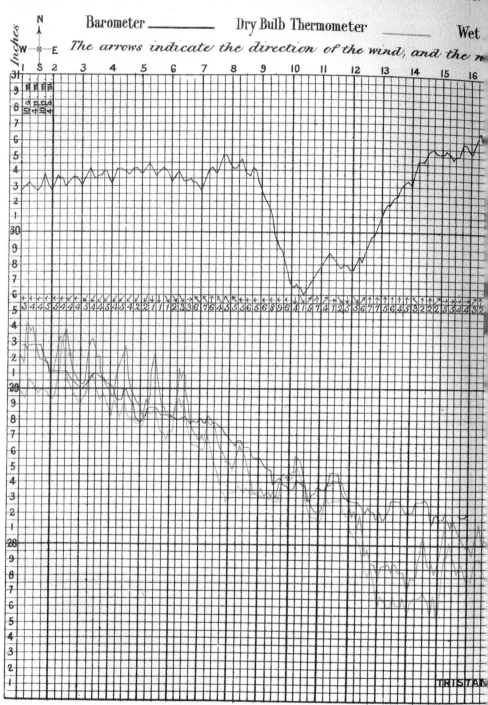

lb Thermometer _____ Temperature of Sea Surface _____

bers beneath its force according to Beaufort's scale

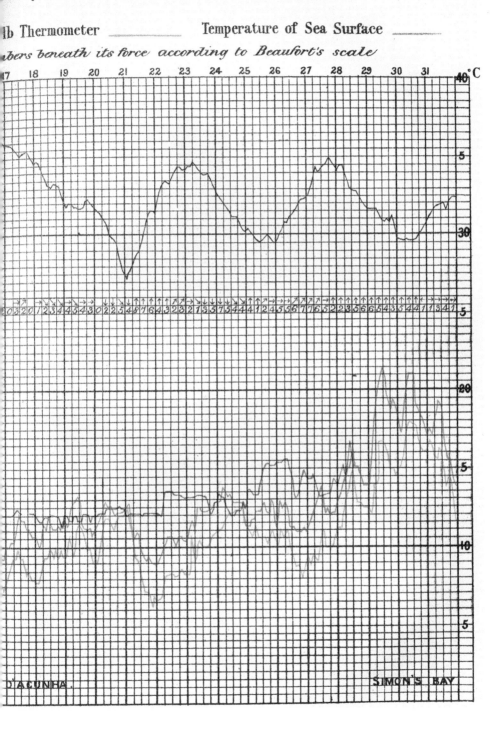

people before resuming the thread of his roving sailor's life.

We landed after breakfast and proceeded to explore the strip along the shore. We were anxious to have reached the plateau, but the sea was breaking heavily on the weather-coast, and it was considered unsafe to land opposite the practicable ascent in a ship's boat. The hut was built to the extreme left of the strip, close to the water-fall, for the convenience of being near the bountiful supply of pure fresh water yielded by the stream. To the right, for about a quarter of a mile, the ground was broken and uneven,—an accumulation of debris from the cliff, covered with a close thicket of well-grown *Phylica arborea* tangled with long grass, and the low branches of the trees overgrown with moss and ferns, the most conspicuous of the latter the handsome *Lomaria robusta,* and the most abundant a spreading *Hymenophyllum* matted over the stones and stumps. The noddy (*Sterna stolida*) builds loose nests of sticks and leaves in the trees, and the ground and the accumulations of moss and dead leaves among the fragments of rock are honey-combed with the burrows of a grey petrel about the size of a pigeon, of a smaller petrel, and of *Prion vittatus.*

The holes of the larger petrel are like rabbit-burrows, and those of the other two birds are smaller. They have the entrances usually more or less concealed, and it is odd to hear the chirping of the birds, old and young, muffled by the layer of soil above them all among one's feet. According to the Stolten-hoffs, the petrels come to the land in large numbers in the beginning of September, having previously been at sea, fishing; when they pair and prepare the burrows for their nests. They disappear for a time in the beginning of October, and return early in November, when the female at once lays a single long-shaped white egg, about the size of a duck's, in the burrow. The young are hatched in the beginning of December. The full-grown bird has a rank taste, which is even communicated to the egg, but the young are good eating. The smaller petrel, a bluish-grey bird, is not much larger than *Thalassidroma wilsoni;* it breeds in company with the *Prion*, in old holes of the larger petrel or in smaller special bur-rows. The smaller petrel and the *Prion* fly chiefly at night or very early morning, and are called, at Tristan, 'night-birds.' The egg of the *Prion* is white, and about the size of a house-pigeon's.

After passing the wood the ground becomes more level, and here the Stoltenhoffs had made a clearing for a potato plot and a vegetable garden. It was a bad season for vegetables, but our blue-jackets carried off a boat-load of cabbages and radishes before the establishment was broken up. They likewise rifled a little hut in the garden, where a large supply of fresh penguins' eggs were stored. Many thrushes and finches were perching on the low trees about, and they were so tame that we had no difficulty in knock-ing down several with our sticks, to get uninjured specimens for stuffing. Both birds are constantly on the island; the thrush builds in the tussock-grass, a couple of feet from the ground, in the beginning of October, and lays usually two eggs—brown spots on a pale greenish ground, very like those of the common blackbird; the finch builds in the bushes and lays

four to five eggs, very like those of the common
canary.

FIG. 37.—Group of rock-hoppers, Inaccessible Island. (*From a photograph.*)

Beyond the garden the tussock-grass of the Tristan
group, which is *Spartina arundinacea*—not *Dactylis
cæspitosa*, the well-known tussock-grass of the Falk-

lands, forms a dense jungle. The root-clumps, or 'tussocks,' are two or three feet in width and about a foot high, and the spaces between them one to two feet wide. The tuft of thick grass-stems—seven or eight feet in height—rises strong and straight for a yard or so, and then the culms separate from one another and mingle with those of the neighbouring tussocks. This makes a brush very difficult to make one's way through, for the heads of grass are closely entangled together on a level with the face and chest. In this scrub one of the crested penguins, probably *Eudyptes chrysocoma*, called by the sealers in common with other species of the genus *Eudyptes* the 'rock-hopper,' has established a rookery. From a great distance, even so far as the hut or the ship, one could hear an incessant noise like the barking of a myriad of dogs in all possible keys, and as we came near the place bands of penguins were seen constantly going and returning between the rookery and the sea. All at once, out at sea a hundred yards or so from the shore, the water is seen in motion, a dark red beak and sometimes a pair of eyes appearing now and then for a moment above the surface. The moving water approaches the shore in a wedge-shape, and with great rapidity a band of perhaps from three to four hundred penguins scramble out upon the stones, at once exchanging the vigorous and graceful movements and attitudes for which they are so remarkable while in the water for helpless and ungainly ones, tumbling over the stones, and apparently with difficulty assuming their normal position upright on their feet—which are set far back—and with their fin-like wings hanging in a useless kind of way at their sides.

When they have got fairly out of the water, beyond the reach of the surf, they stand together for a few minutes drying and dressing themselves and talking loudly, apparently congratulating themselves on their safe landing, and then they scramble in a body over the stony beach—many falling and picking themselves up again with the help of their flappers on the way—and make straight for one particular gangway into the scrub, along which they waddle in regular order up to the rookery. In the meantime a party of about equal number appear from the rookery at the end of another of the paths. When they get out of the grass on to the beach they all stop and talk and look about them, sometimes for three or four minutes. They then with one consent scuttle down over the stones into the water, and long lines of ripple radiating rapidly from their place of departure are the only indications that the birds are speeding out to sea. The tussock-brake, which in Inaccessible Island is perhaps four or five acres in extent, was alive with penguins breeding. The nests are built of the stems and leaves of the *Spartina* in the spaces between the tussocks. They are two or three inches high, with a slight depression for the eggs, and about a foot in diameter. The gangways between the tussocks, along which penguins are constantly passing, are wet and slushy, and the tangled grass, the strong ammoniacal smell, and the deafening noise continually penetrated by loud separate sounds which have a startling resemblance to the human voice, make a walk through the rookery neither easy nor pleasant.

The penguin is thickly covered with the closest

felting of down and feathers, except a longitudinal band, which in the female, extends along the middle line of the lower part of the abdomen, and which, at all events in the breeding season, is without feathers. The bird seats herself almost upright upon the eggs, supported by the feet and the stiff feathers of the tail, the feathers of the abdomen drawn apart and the naked band directly applied to the eggs, doubtless with the object of bringing them into immediate contact with the source of warmth. The female and the male sit by turns; but the featherless space, if present, is not nearly so marked in the male. When they shift sitters they sidle up close together, and the change is made so rapidly that the eggs are scarcely uncovered for a moment. The young, which are hatched in about six weeks, are curious-looking little things covered with black down.

There seems to be little doubt that penguins properly belong to the sea, which they inhabit within moderate distances of the shore, and they only come to the land to breed and moult, and for the young to develop sufficiently to become independent. But all this takes so long that the birds are practically the greater part of their time about the shore. We have seen no reason as yet to question the old notion that their presence is an indication that land is not very far off.

Eudyptes chrysocoma is the only species found in the Tristan d'Acunha group. The males and females are of equal size, but the males may be readily distinguished by their stouter beaks. From the middle of April till the last week in July there

are no penguins on Inaccessible Island. In the end
of July the males begin to come ashore, at first
in twos and threes, and then in larger numbers,
all fat and in the best plumage and condition. They
lie lazily about the shore for a day or two, and then
begin to prepare the nests. The females arrive in
the middle of August, and repair at once to the
tussock-brake. A fortnight later they lay two, rarely
three, eggs, pale blue, very round in shape, and
about the size of a turkey's egg. It is singular
that one of the two eggs is almost constantly
considerably larger than the other. The young are
hatched in six weeks. One or other of the old birds
now spends most of its time at sea, fishing, and the
young are fed, as in most sea-birds, from the crop
of the parents. In December young and old leave
the land, and remain at sea for about a fortnight,
after which the moulting season commences. They
now spread themselves along the shore and about the
cliffs, often climbing, in their uncouth way, into
places which one would have imagined inaccessible
to them. Early in April they all take their departure. The Stoltenhoffs witnessed this exodus
on two occasions, and they say that on both it
took place in a single night. In the evening the
penguins were with them, in the morning they were
gone.

There are three species of albatross on Inaccessible
Island, the wandering albatross, *Diomedea exulans;*
the mollymawk, which appears to be here, *D.
chlororhyncha*, though the name is given by the
sealers to different species—certainly farther south
to *D. melanophrys;* and the piew, *D. fuliginosa.*

About two hundred couples of the wandering albatross visit the island. They arrive and alight singly on the upper plateau early in December, and build a circular nest of grass and clay, about a foot high and two feet or so in diameter, in an open space free from tussock-grass, where the bird has room to expand his wonderful wings and rise into the air. The female lays one egg in the middle of January, about the size of a swan's, white with a band of small brick-red spots round the wider end. The great albatross leaves the island in the month of July.

The mollymawk is a smaller bird, and builds a higher and narrower nest, also usually in the open, but sometimes among the brush and tussocks, in which case it has to make for an open space before it can rise in flight. It breeds a little earlier than the wandering albatross, and its eggs were just in season when we were at Tristan. *Diomedea fuliginosa* builds a low nest on the ledges of the cliffs.

The other common sea-birds on Inaccessible are the sea-hen, here probably *Procellaria gigantea*, which is always on the island, and lays two eggs in October on the ground, and a beautiful delicately-coloured tern, *Sterna cassini*, white and pale grey with a black head and red coral feet and beak, which breeds in holes in the most inaccessible parts of the cliffs.

Inaccessible, like Tristan, has its 'island hen,' and it is one of my few regrets that we found it impossible to get a specimen of it. It is probably a *Gallinula*, but it is certainly a different species from the Tristan bird. It is only about a fourth the size, and it seems to be markedly different in appearance. The Stoltenhoffs were very familiar with it, and described it as being exactly like a black chicken two days old, the legs and beak black, the beak long and slender, the head small, the wings short and soft and useless for flight. It is common on the plateau, and runs like a partridge among the long grass and ferns, feeding upon insects and seeds. An 'island hen' is also found on Gough Island; but the sealers think it is the same as the Tristan species.

Some of our party returned to the ship about mid-day, and we cruised round the island, the surveyors plotting in the coast-line, and thus filling up a geographical blank, and in the afternoon we dredged in sixty and seventy-five fathoms.

We returned to the anchorage about seven o'clock, and the exploring parties came on board, the Germans accompanying them with all their gear. As we hove in sight of the hut a broad blaze shot up, followed by a dense volume of smoke, and in a few minutes the solitary human habitation on Inaccessible Island was reduced to a heap of ashes. I do not know whether the match was put to the dry straw of the thatch by accident or by design, but the Stoltenhoffs seemed to feel little regret at the destruction of their dwelling. They left the place with no very friendly feelings towards their Tristan neighbours, and had no wish to leave anything behind them which might be turned to their use.

Early on the morning of Friday, the 17th, we were off Nightingale Island, so named after the Dutch skipper who first reported it. The outline of this island is more varied than that of the other two, and its geological structure is somewhat different. Towards the north end there is a conical peak of a grey, rudely columnar basaltic rock 1,105 feet high, and the southern portion of the island, which is more undulating, consists of bedded tufts with included angular fragments of dolerite, like the rocks above the settlement in Tristan. Near the south shore these softer rocks run up into a second lower ridge, and a low cliff bounds the island twenty or thirty feet high, with creeks here and there where boats can land through the surf. In the sea-cliff there are some large caves worn in the friable rock, which used to be the favourite haunts of the fur-seal and the sea-elephant; but these have been nearly exterminated, and the annual visit of the sealers from Tristan is rapidly reducing the small number which still come to the island in the pupping season.

The ship stopped off the east end of the island to land surveying and exploring parties at the foot of what looked at a distance like a gentle slope of meadow with some thickets of low trees, running up into the middle of the island, between the two elevations.

The party who landed found, however, that instead of a meadow the slope was a thick copse of tussock-grass,—and one mass of penguins. Struggling through the dense matted grass which reached above their heads, they could not see where they were going, and they could not move a step without crushing eggs or old or young birds. The crowds of penguins resenting the intrusion with all the

PLATE XXXI.—NIGHTINGALE ISLAND.

vigour at their command, yelled and groaned and
scrambled after their legs, and bit and pecked them
with their strong sharp beaks till the blood came.
What with the difficulty of forcing their way through
the scrub, the impossibility of seeing a foot before
them in the grass, the terrific noise which prevented
shouts being heard, and the extraordinary sensation
of being attacked about the legs by legions of
invisible and unfamiliar enemies, some of the servants
got nervous and bewildered. They lost their own
masters, and were glad to join and stick to any
one whom they were fortunate enough to find, and
thus several of our explorers got separated from their
apparatus, and some lost their luncheons.

Fortunately at five o'clock all our party returned
in safety to the ship, save one,—a fine old setter
answering to the name of 'Boss,' one of a brace
we had on board for sporting purposes, got astray
among the penguins. His voice, clamorous for a
time in his bewilderment and fear and the torture
he endured from the beaks of the penguins, was
soon lost in the infernal uproar; and as the men
had enough to do to look after their own safety
they were compelled reluctantly to leave him to his
fate.

Since our visit the remote little community of
Tristan d'Acunha has not entirely escaped political
complications, such as have involved many States
of greater importance in their own estimation. The
attention of the Lords of the Admiralty was for
some reason or other attracted to the Island, and
H. M. S. 'Sappho,' commander Noel Digby, called
at Tristan in January, 1875, and Captain Digby

reported that at that time there were fourteen families on the Island, eighty-five persons in all. The condition of the islanders seem to have been much the same as when we went there two years before. From Captain Digby's report, it appeared to Lord Carnarvon that if the Tristan group really formed part of the Cape Colony, which seemed to be the case from Bishop Gray having visited it as part of his diocese, the jurisdiction of the Cape Government should be recognised; and it might be well that certain limited magisterial powers should be conferred upon Peter Green and perhaps one or two others by the Governor; and he wrote to Sir Henry Barkly for information and suggestions. Sir Henry Barkly replied that Tristan d'Acunha had certainly been included in the letters patent constituting the see of Capetown, but that on the creation of the bishoprick of St. Helena it had been transferred to that diocese, and that no proclamation or other evidence could be found giving the Governor of Cape Colony special authority over the Tristan group. Moreover Sir Henry Barkly's advisers reported that in the present state of information relating to the connection between Tristan d'Acunha and the Cape Colony they did not feel at liberty to recommend that magisterial powers should be conferred on any of the inhabitants by the Cape Government.

Lord Carnarvon then gave up the idea of attaching Tristan to the Cape, and proposed that its government should be provided for under certain rules, such as exist in the case of Norfolk Island; one or two magistrates being appointed with authority to settle small disputes, celebrate marriages, and look to the maintenance of order; it was provided that the chief magistrate should communicate, as occasion occurred, with the Secretary of State for the Colonies, and any graver matters would doubtless be adjusted by him by giving special powers to the Captain of one of H M.'s ships. The Secretary for the Colonies suggested that the Admiralty should direct the officer in command of a ship likely to call at the Island to appoint one or two magistrates and to confer with them as to the rules for their government; and accordingly in October, 1875, H. M. S. 'Diamond' visited Tristan, and Captain Stanley Bosanquet forwarded a very full and entertaining report to the Admiralty. He said that if he had failed in carrying out the wishes of H. M.'s Secretary for the Colonies it was because on becoming acquainted with the settlers he was unable to see any need of establishing rules for their future guidance; he again took a census of the population which remained stationary at eighty-five, and it appears that there are now only fifteen males of the age of twenty-one years and upwards. "These," he remarks, "represent the physical force, and I may also say the intellectual, of this somewhat unsophisticated community, although I should not venture to assert this (superiority of the males) of any more highly civilized one," and from what I saw of the business capacity of the Tristan young ladies and their excellent physical development I should certainly have thought twice before venturing to assert it even there,—"the families are connected by the ties of marriage, and their interests are identical. They have certain rules of their own, and the present senior male member of the community, Peter Green, is made their referee if necessary."

Captain Bosanquet doubts the necessity of the emigration of any of the settlers even if the population increase considerably, he thinks that there is land and stock enough for a much larger number. He says, " I have little doubt that the peculiar enjoyment and content of the original few settlers has now to a great extent diminished. It depended upon ample space, and abundance, and undisturbed possession ; also this and the neighbouring islands and sea abounded with seals, sea-elephants, and wild goats, which were easily taken and in very great numbers ; and there was an extensive traffic for the few with the whale-ships which then constantly communicated. With the increase of the inhabitants, however, their unbounded freedom was curtailed, as there were more people who had claims to be respected ; there were more mouths to feed, and more hands to take part in the seal-hunting, &c., and to share in the traffic ensuing therefrom."

The loss of the traffic with the whalers, and the consequent scarcity of foreign productions, is certainly the great difficulty of the Tristaners, but I doubt if even that is so great as they represent ; as I have already said from eight to twelve ships still call yearly, and as all of them are in want of fresh provisions, and the islanders are very shrewd at a bargain, they probably might easily get all they require. They seemed to us to be fully alive to the advantage of making the worst of things. Notwithstanding his satisfaction with the existing state of matters, Captain

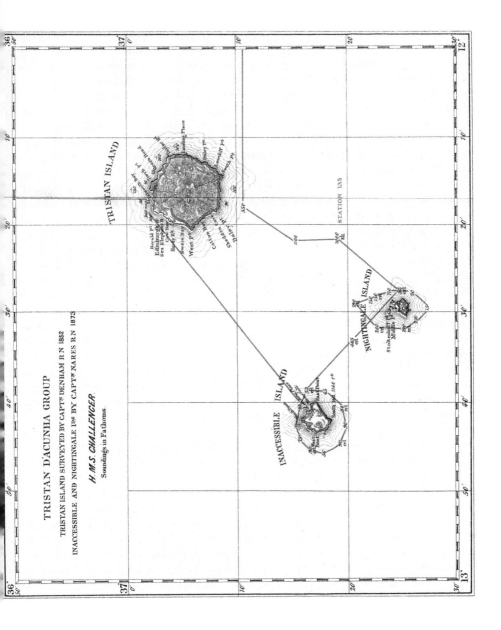

TRISTAN D'ACUNHA GROUP

TRISTAN ISLAND SURVEYED BY CAPT.ⁿ DENHAM R.N 1852
INACCESSIBLE AND NIGHTINGALE Iᵈˢ BY CAPT.ˢ NARES R.N 1873

H.M.S. CHALLENGER.
Soundings in Fathoms.

TRISTAN ISLAND

INACCESSIBLE ISLAND

NIGHTINGALE ISLAND

STATION 135

The material originally positioned here is too large for reproduction in this
reissue. A PDF can be downloaded from the web address given on page iv
of this book, by clicking on 'Resources Available'.

Bosanquet makes some suggestions, most of which have been concurred in by the Secretary for the Colonies, although the most important, namely that the naval officer in command of the South African State should be *ex officio* governor of the island, seemed open to so many objections that it has not been adopted.

A proposal of Lord Carnarvon's, to give two hundred pounds worth of useful presents to the islanders of things which they cannot easily obtain from passing ships, will no doubt be highly popular. They had, it seems, represented that a clergyman was one of their most urgent needs ; an educated man, clerical or lay, of a certain stamp among them would be an enormous advantage ; but an educated man of another stamp, such as they were much more likely to get, would be very much the reverse.

My own impression is that it would have been just as well to have left the settlers of Tristan d'Acunha alone. At present there is a general feeling of equality, and their arbiter is of their own choosing ; and they took special care that it should be fully understood that their deference to Peter Green was purely voluntary. I should fear that the appointment of magistrates from among themselves by external authority may give rise to all kinds of jealousy and ill will. If the place is understood to belong to Great Britain at all, it is no doubt important that in such a case as that of the 'Shenandoah,' they should be able to produce evidence to that effect. The Tristaners of the present day have certainly not left the most favourable impression on my mind. They are by no means ill off ; they are very shrewd and

sufficiently greedy ; and their conduct to the Stoltenhoffs, if their story be true, which we have never had
any reason to doubt, in landing surreptitiously and
killing the last of the flock of goats on Inaccessible
Island, if not actually criminal, was to say the least
most questionable.

While the party on land were struggling among the
tussocks and penguins, and gaining an experience of
the vigour of spontaneous life, animal and vegetable,
which they are not likely soon to forget, the ship
took a cruise round the island to enable the surveyors to put in the coast line ; and in the afternoon
the hauls of the dredge were taken in 100 and 150
fathoms. A large quantity of things were procured
of all groups, the most prominent a fine species of
Primnoa, many highly-coloured *Gorgoniæ*, and a
very elegant *Mopsea* or some closely allied form.
Lophohelia prolifera or a very similar species was
abundant, associated with an *Amphihelia* and a fine
Cœnocyathus. Hydroids and sponges were in considerable number tangled in masses with calcareous
and horny bryozoa. There were a few star-fishes, and
a very few mollusca. The whole assemblage resem-
bled a good deal the produce of a haul in shallow
water off the Mediterranean coast of Morocco.

On the following day we crossed over, sounding and
dredging on our way, to within a mile and a half of
the west shore of the Island of Tristan. A haul in
1,000 fathoms gave us somewhat to our surpise some
most typical samples of the common *Echinus flemingii*,
along with *Ophiomusium lymani* and *Rhizocrinus*.
We dredged a second time as we were nearing Tristan
in 550 fathoms, and took one or two specimens of

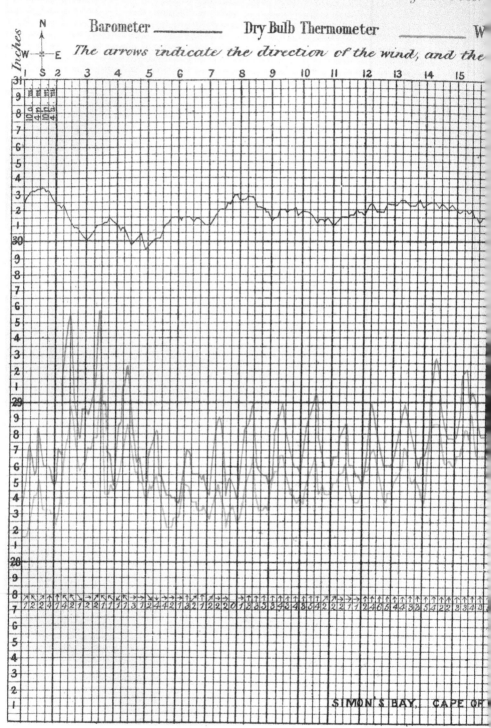

PLATE XXXIII. *Meteorological Obser*

Barometer ——————— Dry Bulb Thermometer —————— W

The arrows indicate the direction of the wind, and the

SIMON'S BAY, CAPE OF

Bulb Thermometer   Temperature of Sea Surface

umbers beneath its force according to Beaufort's scale

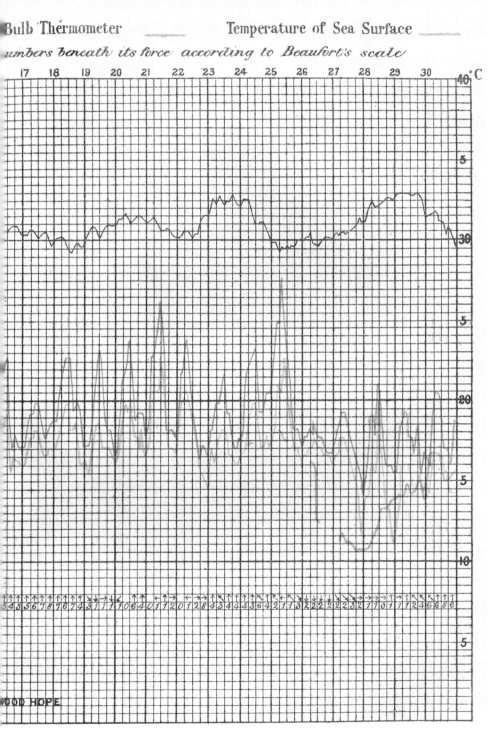

54356787674311110640172012843444364217132222223271131712466886

OOD HOPE

a species of *Antedon,* some corals the same as those off Nightingale Island, and a quantity of bryozoa. In the evening we set all plain sail, and, with a favouring breeze from the north-west, proceeded on our voyage towards the Cape.

On the 20th of October we sounded in 2,100 fathoms, on hard ground, with a bottom temperature of $1°·1$ C., and put the dredge over. The dredge got entangled at the bottom, and was disengaged with some difficulty; it came up in the evening quite empty, and we made all plain sail and proceeded on our way.

On the 21st the weather changed, the wind dropping round to the S.S.E., and blowing very cold with force = 8. We ran on under double-reefed topsails, and during that day and the next made nearly 350 miles.

On the 23rd we put the' dredge over early, and veered to 3,000 fathoms; and after breakfast we sounded in 2,550 fathoms, with a bottom of reddish clay and a bottom temperature of $0°·7$ C., the position of the sounding being lat. 35° 59′ S., long. 1° 25′ E. The dredge contained little save some scarlet caridid and peneid shrimps; but two small star-fishes gave evidence that it had reached the bottom.

On the 25th we sounded in 2,650 fathoms, with a bottom of reddish clay, and a bottom temperature of $1°·0$ C.; and on the 27th in 2,325 fathoms, with a bottom of grey ooze with nodules of black manganese, and a bottom temperature of $0°·47$ C. The distance from the Cape of Good Hope at noon was 138 miles. A series of temperature soundings was taken at intervals of ten fathoms from the surface to a depth of a hundred fathoms with the following results :—

Surface	13° 4 C.	60 fathoms . . .	11° 2
10 fathoms . . .	13 · 4	70 ,, . . .	11 · 1
20 ,, . . .	13 · 3	80 ,, . . .	11 . 1
30 ,, . . .	12 · 8	90 ,, . . .	11 · 0
40 ,, . . .	11 · 6	100 ,, . . .	10 · 9
50 ,, . . .	11 · 3		

On the 28th we stopped at 7 a.m., and sounded in 1,250 fathoms with a bottom of grey mud, Table Mountain and the range of hills above Simon Bay being fairly visible on the north-eastern horizon. We took a series of temperatures at intervals of ten fathoms down to a hundred. At noon we took a second series, fifteen nautical miles to the south-west of the Cape of Good Hope, and we found that in the interval we had passed into the loop of the Agulhas current, which curls round the Cape close to the land. The contrast between the two series is remarkable.

	7 A.M.	Noon.
Surface ...	14° 6 C.	16° 7 C.
10 fathoms.	14 · 7	17 · 1
20 ,,	14 · 5	16 · 8
30 ,,	14 · 4	16 · 4
40 ,,	13 · 8	15 · 8
50 ,,	12 · 5	14 · 7
60 ,,	12 · 3	13 · 9
70 ,,	11 · 6	13 · 3
80 ,,	11 · 6	12 · 8
90 ,,	11 · 7	12 · 2
100 ,,	11 · 6	11 · 0

The temperature of the air likewise rose perceptibly, the thermometer in the shade indicating at noon

15° C., nearly three degrees above the average of the same hour during the previous week.

At two o'clock p.m. we rounded the Cape, and signalled our number to H.M.S. 'Rattlesnake,' just returned from the scene of the Ashantee War; and an hour later we cast anchor in Simon's Bay, and bade farewell for many a long day to the friendly waters of the Atlantic.

PENGUINS AT HOME.

APPENDIX A.

Table of Temperatures observed between Bahia and the Cape of Good Hope.

Depth in Fathoms.	Station No. 129. Lat. 20° 13′ S. Long. 35° 19′ W.	Station No. 130. Lat. 26° 15′ S. Long. 32° 56′ W.	Station No. 131. Lat. 29° 35′ S. Long. 28° 9′ W.	Station No. 132. Lat. 35° 25′ S. Long. 23° 40′ W.	Station No. 133. Lat. 35° 41′ S. Long. 20° 55′ W.	Station No. 134. Lat. 36° 12′ S. Long. 12° 16′ W.
Surface.	23°· 3 C.	20°· 6 C.	18°· 3 C.	14°· 4 C.	13°· 9 C.	12°· 8 C.
25	...	...	...	. .	13 · 9	...
50	...	...	...	...	13 · 9	...
75	19 · 0	...	...	...	13 · 4	...
100	17 · 3	15 · 0	16 · 0	12 · 9	13 · 0	...
200	11 · 0	12 · 1	11 · 6	10 · 0	...	7 · 7
300	6 · 6	7 · 5	9 · 2	6 · 4	...	...
400	4 · 2	5 · 1	5 · 4	4 · 3	...	...
500	3 · 5	3 · 7	4 · 0	3 · 6	...	...
600	...	2 · 4	2 · 8	...	...	...
700	2 · 6	2 · 8	2 · 8	...	...	2 · 6
800	2 · 8	2 · 6	...	...	...	...
900	2 · 9	2 · 5	2 · 7	...	...	...
1000	2 · 3	2 · 5	2 · 5	...	...	...
1100	...	2 · 0	...	...	...	2 · 3
1200	2 · 8	2 · 4	...	...	...	...
1300	1 · 7	2 · 3	...	...	...	2 · 4
1400	1 · 9	2 · 2	...	...	...	...
1500	2 · 2	2 · 6	...	...	...	...
Bottom Temperature.	0°· 6	0°· 8	0°· 7	1°· 1	1°· 3	1°· 6
Depth.	2150	2350	2275	2050	1900	2025

Depth in Fathoms.	Station No. 135. Lat. 37° 20′ S. Long. 12° 25′ W.	Station No. 136. Lat. 36° 43′ S. Long. 7° 13′ W.	Station No. 137. Lat. 35° 59′ S. Long. 1° 34′ E.	Station No. 138. Lat. 36° 22′ S. Long. 8° 12′ E.	Station No. 139. Lat. 35° 35′ S. Long. 16° 8′ E.	Station No. 140. Lat. 35° 0′ S. Long. 7° 57′ W.
Surface.	12°· 0 C.	12°· 2 C.	13°· 4 C.	13°· 4 C.	13°· 4 C.	14°· 6 C.
25	...	...	...	...	...	...
50	...	...	13 · 3	13 · 4	11 · 3	12 · 5
75	...	...	...	...	...	11 · 6
100	9 · 1	11 · 2	13 · 0	13 · 3	10 · 7	...
200	...	9 · 5	11 · 0	10 · 4	7 · 7	...
300	5 · 8	6 · 2	6 · 8	6 · 7	4 · 9	...
400	3 · 9	4 · 0	4 · 3	4 · 3	3 · 1	...
500	3 · 4	3 · 4	3 · 5	3 · 3	2 · 9	...
600	2 · 6	2 · 7	...	...	...	...
700	...	...	2 · 6	...	2 · 6	...
800	...	...	...	...	...	...
900	...	...	2 · 5	...	...	...
1000	...	...	...	...	...	...
1100	...	...	...	...	2 · 3	...
1200	...	...	...	...	...	...
1300	..	...	2 · 2	...	2 · 2	...
1400	...	...	...	..	...	...
1500	...	...	2 · 0	...	2 · 0	...
Bottom Temperature.	...	1°· 1	0°· 7	1°· 0	0°· 5	...
Depth.	...	2100	2550	2650	2325	...

APPENDIX B.

Table of Serial Soundings down to 200 fathoms, taken between Bahia and the Cape of Good Hope.

Depth in Fathoms.	Station No. 137. Lat. 35° 59′ S. Long. 1° 34′ E.	Station No. 138. Lat. 36° 22′ S. Long. 8° 12′ E.	Station No. 139. Lat. 35° 35′ S. Long. 16° 8′ E.	Depth in Fathoms.	Station No. 139. Lat. 35° 35′ S. Long. 16° 8′ E.	Station No. 140. Lat. 35° 0′ E. Long. 17° 57′ E.	Station No. 140.
Surface.	13°· 4 C.	13°· 4 C.	13°· 4 C.	Surface.	13°· 4 C.	14°· 6 C.	16°. 7 C.
50	13 · 3	13 · 4	11 · 3	10	13 · 4	14 · 7	17 · 1
100	13 · 0	13 · 3	10 · 7	20	13 · 3	14 · 5	16 · 8
150	12 · 3	11 · 7	8 · 4	30	12 · 8	14 · 4	16 · 4
200	11 · 0	10 · 4	7 · 7	40	11 · 6	13 · 8	15 · 8
250	8 · 7	8 · 2	6 · 6	50	11 · 3	12 · 5	14 · 7
300	6 . 8	6 · 7	4 · 9	60	11 · 2	12 · 3	13 · 9
350	5 · 7	5 · 2	3 · 6	70	11 · 1	11 · 6	13 · 3
400	4 · 3	4 · 3	3 · 1	80	11 · 1	11 · 6	12 · 8
450	4 · 0	3 · 7	3 · 0	90	11 · 0	11 · 7	12 · 2
500	3 · 5	3 · 3	2 · 9	100	10 · 9	11 · 6	11 · 0

APPENDIX C.

Table of Specific Gravity observations taken between Bahia and the Cape of Good Hope during the Months of September and October, 1873.

Date 1873.	Latitude S.	Longitude W.	Depth of the Sea.	Depth (δ) at which the water was taken.	Temperature (t) at δ.	Temperature (t') during observation.	Specific Gravity at (t). Water at 4° = 1.	Specific Gravity at 15°·5. Water at 4° = 1.	Specific Gravity at (t). Water at 4° = 1.
			Fms.	Fms.					
Sept. 26	13° 45'	37° 59'		Surface.	25°·1C.	25°·0C.	1·02514	1·02776	1·02512
27	14 51	37 1		Surface.	25·3	25·6	1·02487	1·02770	1·02498
28	17 7	36 50		Surface.	24·7	24·9	1·02517	1·02775	1·02520
29	19 6	35 40		Surface.	23·6	24·0	1·02530	1·02761	1·02550
30	20 13	35 19	2150	Surface.	23·4	24·3	1·02518	1·02757	1·02546
,,	...	...		50	...	23·9	1·02509	1·02736	·...
,,	...	...		100	17·3	23·9	1·02509	1·02736	1·02690
,,	...	...		200	11·0	23·8	...	1·02736	1·02834
,,	...	...		300	6·6	24·8	1·02328	1·02558	1·02712
,,	...	...		400	4·2	23·9	...	1·02558	1·02741
Oct. 1	22 15	35 37		Surface.	22·8	22·9	1·02547	1·02774	1·02548
2	24 43	34 17		Surface.	21·0	21·5	1·02562	1·02717	1·02574
3	26 15	32 56		Surface.	21·0	21·6	1·02546	1·02703	1·0..560
,,	...	...		4	...	21·4	1·02552	1·02704	...
,,	...	...		50	...	21·7	1·02523	1·02682	
,,	...	...		100	15·0	21·5	1·02496	1·02649	1·02660
,,	...	...		200	12·1	21·5	1·02454	1·02608	1·02679
,,	...	...		300	7·5	21·6	1·02416	1·02573	1·02719
,,	...	...		400	5·1	21·5	1·02402	1·02554	1·02727
,,	...	...'	2350	Bottom.	0·8	21·5	1·02552	1·02706	1·02916?
4	27 43	31 3		Surface.	19·4	20·1	1·02588	1·02702	1·02603
5	29 1	28 59		Surface.	18·9	19·4	1·02573	1·02690	1·02601
6	29 35	28 9	2275	Surface.	18·3	19·1	1·02575	1·02665	1·02593
,,	...	...		100	16·0	18·5	1·02555	1·02629	1·02616
,,	...	...		200	11·6	18·5	1·02528	1·02602	1·02683
,,	...	...		300	9·2	18·6	1·02489	1·02565	1·02687
,,	...	...		400	5·4	18·7	1·02462	1·02540	1·02715
,,	...	...		1000	2·5	18·7	1·02494	1·02572	1·02767
7	29 11	26 25		Surface.	18·3	18·7	1·02576	1·02654	1·02581
8	31 22	26 54		Surface.	16·6	16·8	1·02608	1·02638	1·02610
9	33 57	24 33		Surface.	14·8	15·4	1·02652	1·02648	1·02660
10	35 25	23 40		Surface.	14·6	15·3	1·02615	1·02609	1·02630
,,	...	...		100	13·3	15·9	1·02605	1·02613	1·02659
,,	...	...		200	10·2	15·6	1·02570	1 02571	1·02676
,,	...	...		300	6·4	16·0	1·02582	1·02543	1·02703
,,	...	...		400	4·2	16·0	1·02535	1·02541	1·02725
,,	...	...	2050	Bottom.	1·1	15·9	1·02572	1·02580	1·02784
11	35 41	20 55		Surface.	13·9	15·2	1·02624	1·02617	1·02648
,,	...	...	1900	Bottom.	1·3	14·6	1·02598	1·02577	1·02786
12	36 10	17 52		Surface.	12·7	13·2	1·02640	1·02590	1·02648
13	36 7	14 27		Surface.	12·0	12·3	1·02653	1·02585	1·02658
14	36 12	12 18		Surface.	12·8	13·0	1·02660	1·02606	1·02661
,,	...	...	2025	Bottom.	1·6	11·5	1·02656	1·02573	1·02775
19	37 5	9 40		Surface.	12·0	12·5	1·02676	1·02612	1·02684

Date 1873.	Lati-tude S.	Longi-tude W.	Depth of the Sea.	Depth (δ) at which the water was taken.	Temperature (t) at δ.	Temperature (t') during observation.	Specific Gravity at (t'). Water at $4° = 1$.	Specific Gravity at 15°·5. Water at $4° = 1$.	Specific Gravity at (t). Water at $4° = 1$.
				Fms.					
Oct. 20	36° 43′	7° 13′		Surface.	12°· 2C.	13°· 0C.	1·02660	1·02606	1·02677
,,	...	...		100	11 · 2	13 · 9	1·02623	1·02586	1·02676
,,	...	...		200	9 · 5	13 · 9	1·02605	1·02568	1·02685
,,	...	...		300	6 · 2	13 · 9	1·02579	1·02542	1·02703
,,	...	...		400	4 · 0	13 · 8	1·02574	1·02535	1·02722
,,	...	...	2100	Bottom.	1 · 1	14 · 0	1·02615	1·02580	1·02784
22	35 57	0 15		Surface.	13 · 6	13 · 8	1·02650	1·02611	1.02650
23	35 59	1 26E		Surface	13 · 4	13 · 6	1 02669	1 02626	1·02669
,,	...	...		100	13 · 0	13 · 2	1·02645	1·02594	1·02646
,,	...	...		200	11 · 0	13 · 2	1·02635	1·02584	1·02677
,,	...	...		300	6 · 8	13 · 4	1·02605	1·02558	1·02711
,,	...	...		400	4 · 3	13 · 0	1·02600	1·02547	1·02729
24	36 2	5 27	2550	Bottom.	0 · 7	12 · 8	1·02633	1·02574	1·02782
25	36 22	8 12		Surface.	12 · 2	13 · 0	1·02640	1·02586	1·02656
,,	...	...		Surface.	15 · 0	15 · 2	1·02629	1·02621	1·02630
26	35 59	11 43	2650	Bottom.	1 · 0	13 · 5	1·02614	1·02570	1·02777
			10.30A.M	Surface.	15 · 6	15 · 6·	1·02653	1·02654	1·02654
,,	...	...	4·30P.M	Surface.	14 · 6	14 · 7	1·02669	1·02650	1·02670
27	35 35	16 8	8·0 P.M.	Surface.	13 · 3	12 · 9	1·02668	1·02613	1·02660
				Surface.	13 · 7	13 · 7	1·02644	1·02603	1·02644
28	35 0	17 57	2325	Bottom.	0·47	14 · 0	1·02605	1·02571	1·02780
,,	...	...	11.30A.M	Surface.	16 · 7	17 · 1	1·02577	1·02615	1·02583
,,	...	...	2.5 P.M.	Surface.	15 · 0	15 · 9	1·02602	1·02610	1·02622
,,	...	...		20	16 · 8	15 · 6	1·02614	1·02615	1·02581
,,	...	...		50	14 · 7	15 · 3	1 02612	1·02605	1·02622

CHAPTER IV.

THE VOYAGE HOME.

ON the morning of the 20th of January, 1876, the 'Challenger' passed through the 'first narrows' of the Strait of Magellan, wind and tide in her favour, at the rate of about seventeen knots an hour; shortly after mid-day she rounded Cape Virgins, and a long uneasy swell gave us somewhat unpleasant evidence of the most welcome fact that we were once more yielding to the pulses of the broad Atlantic.

For the previous three weeks we had been creeping down inside the islands from the Gulf of Penas, through the Messier and the Sarmiento Channels and the Magellan Strait, sounding and trawling nearly every day; and we had amassed a splendid series of characteristic Patagonian forms from depths of 60 to 400 fathoms. On the afternoon of the 20th we sounded in 55 fathoms, about 20 miles due east of Cape Virgins, with a bottom of blackish sand and a bottom temperature of 8°·8 C.

The trawl brought up a large number of a wonderfully handsome *Euryale,* the disk in some of the specimens between three and four inches across. We put a number of these great disks into absolute alcohol to harden the tissues at once and preserve them in the best condition for dissection. There were also some very large simple Ascidians (*Cynthia gigas*) from 30 to 40 centims. long, and with the ganglion— usually a minute body not at once detected, lying between the two orifices—a well-defined grey mass nearly as large as a pea. A viviparous ophiurid occurred in considerable numbers, which we had already found in shallow water off Kerguelen Island. I shall give an account of its singular mode of reproduction when describing the shallow water dredgings at the Falkland Islands, in which it occurred plentifully. On the following day we trawled in 70 fathoms about midway between Cape Virgins and the Jason Islands. Animals were still abundant, but most of them of known forms. A pretty little *Chirodota* which adhered in numbers to the meshes of the trawl was perhaps the most interesting on account of its unusually large and numerous wheels. The bottom was a black sand, and the bottom temperature 7°.8 C.

The 22nd was a wretched day, with cold rain and fog and a disagreeable swell; we sounded in the morning in 110 fathoms and put over the trawl, but it came up empty; owing to a strong current setting northward it had probably never reached the bottom. We had hoped to have reached Stanley Harbour before night, but during most of the day the fog was so thick that it was unsafe to run towards the land. In the afternoon we sighted the Jason Islands, and in the evening it cleared up and we had a good view of the little group,—Jason West, Jason East, Grand Jason, Steeple Jason, and Elephant Jason, rocky islets rising abruptly from the sea. We had a fine run during the night along the north coast of the Falklands; at half-past five next morning Cape Bougainville was seen due south of us. The weather was showery and squally, with a strong southerly breeze, but the land became more distinct during the forenoon as we passed the entrance of Berkeley Sound, and some rather high hills could be seen at intervals between the showers. At two o'clock we passed Pembroke lighthouse, and slipped quietly between the headlands into the little land-locked bay which forms the harbour of Stanley, the present seat of government of the Falkland Islands.

At a first glance these islands are not attractive, and I doubt if they improve greatly on acquaintance.

The land is generally low and flat, but it rises here and there into ridges, the highest a little over 2,000 feet in height. The ground is dark in colour, a mixture of brown and dull green; the ridges are pale grey with lines of outcrop of hard white quartzite, like dilapidated stone walls, at different levels along the strike. The vegetation is scanty, and what little there is, very ineffective. There is nothing of higher dignity than a herb, the nearest approach to a shrub being a rank form of groundsel (*Senecio candicans*), with large button-like yellow flowers and very white woolly foliage, which runs up along the shore and in sheltered nooks inland to a height of two or three feet; and a pretty *Veronica* (*V. decussata*), which is however indigenous on the west island only, and is introduced in the gardens about Port Stanley.

Above Stanley Harbour the land slopes up for a hundred feet or so to a low ridge, beyond which what is called there the 'Camp' (*champ*) extends nearly level for many miles, with slightly raised stretches of pasture and wide patches of peat and dark boggy tarns. The little town of Stanley is built along the shore and stretches a little way up the slope. It is built mainly of square, white, grey-slated houses, and puts one greatly in mind of one of the newer small towns in the Scottish West Highlands or in one of the Hebrides. The resemblance is heightened by the smell of peat-smoke, for peat is almost universally burned as there is no wood and coal costs 3*l.* a ton. The Government House is very like a Shetland or Orkney manse, stone-built, slated, and grey, without the least shelter. In the square grass paddock surrounded by a low wall, stretching from the house

to the shore, a very ornamental flock of Upland
geese were standing and preening their feathers the
first time we called there. This tameness of the sea-
birds is still most remarkable in the Falkland Islands,
and a strange contrast to their extreme wildness in
the Strait of Magellan; there we stalked the kelp
goose (*Chloephaga antarctica*), and the steamer duck
(*Micropterus cinereus*), day after day with great labour
and but little success, finding great difficulty in
getting even within long range of them; while in the
Falklands the same species were all about, standing
on the shore within stone-throw, or diving or fishing
quietly within a few yards of the boats. I was told
that they are not now nearly so tame, however, as they
were some years ago. Almost every evening we met
some one coming to the settlement with a string of
Upland geese for the pot, and I suppose it is beginning to dawn upon the poor birds that their new
neighbours are not so harmless as they look. Very
likely it may take some generations of experience to
make them thoroughly wary, and the difference
between the birds of the Islands and those of the
Straits may probably be, that while the former have
been safe in their primeval solitude up to within a
recent period, the latter have been selecting themselves for ages on their capacity for eluding the craft
of hungry Patagonians and Fuegians.

The town is clean and well kept, and even the
smallest houses are tidy and have a well-to-do look.
Many of the houses belonging to the agents of the
Falkland Islands Company and to the representatives
of several private firms have very pretty green-houses
attached to them, the gay groups of fuchsias and

pelargoniums of all the best home varieties contrasting pleasantly with the desolation outside. The Government Barrack, occupied by an officer and a company of marines, is rather an imposing structure with a square tower in the middle of the town; and there is a neat little Episcopal Church.

The Falkland Islands were first seen by Davis in the year 1592, and Sir Richard Hawkins sailed along their north shore in 1594. In 1598 Sebald de Wert, a Dutchman, visited them, and called them the Sebald Islands, a name which they still bear on some of the Dutch maps. Captain Strong sailed through between the two principal islands in 1690, and called the passage Falkland Sound. In 1763 the islands were taken possession of by the French, who established a colony at Port Louis; they were however expelled by the Spaniards in 1767 or 1768. In 1761 Commodore Byron took possession on the part of England on the ground of prior discovery, and his doing so was nearly the cause of a war between England and Spain, both countries having armed fleets to contest the barren sovereignty. In 1771, however, Spain yielded the islands to Great Britain by convention. Not having been actually colonized by us, the Republic of Buenos Ayres claimed the islands in 1820, and formed a settlement at the old Port Louis, which promised to be fairly successful, but owing to some misunderstanding with the Americans it was destroyed by the latter in 1831. After all these vicissitudes the British flag was once more hoisted at Port Louis in 1833, and since that time the Falkland Islands have been a regular British colony under a Governor. The group was called by the French the

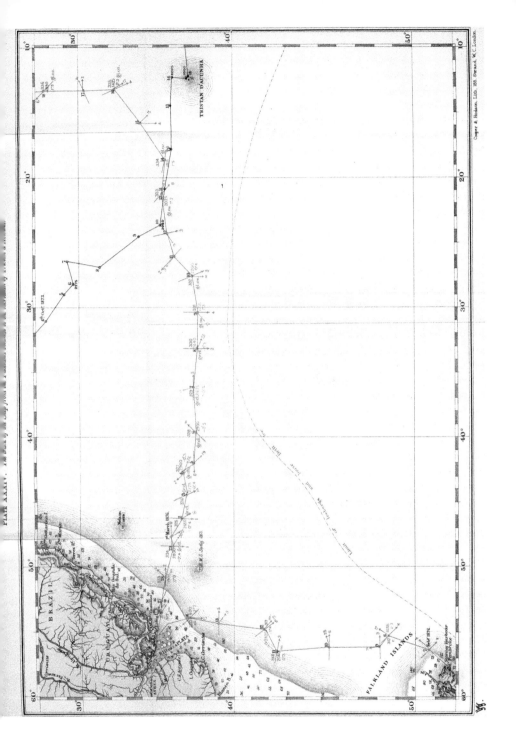

The material originally positioned here is too large for reproduction in this reissue. A PDF can be downloaded from the web address given on page iv of this book, by clicking on 'Resources Available'.

Malouines, from the inhabitants of St. Maloes whom they imagine to have been their first discoverers; and the Spanish name, the Malvinas, the most euphonious of them all, is the one still mostly in use by their neighbours of South America.

The islands are about a hundred in number, but only two of them are of any size. They lie between the parallels of 51° and 52° 45′ South, and the meridians of 57° 20′ and 61° 46 West. The climate is very miserable, considering that the latitude corresponds with that of Middlesex; for though the thermometer rarely falls in winter much below the freezing point, it rarely rises in summer much above 18°·5 C.; and fog and rain are so constant and sunshine so scarce, that wheat will not ripen, barley and oats can scarcely be said to do so, and the common English vegetables will not produce seed in the gardens. Still the colony appears to be very healthy, the inhabitants seem to get thoroughly accustomed to their moist, chilly surroundings, and the only 'pale maidens' to be seen are the drooping delicate flowers of *Sisyrinchium filifolium*, which cover the 'camp' round Stanley in early spring and have earned that pretty soubriquet. Of late years the industry of the Falkland Islands has been developing most rapidly. It has been found that the pasture is even more suitable for sheep than for cattle; and in 1872 the Falkland Islands Company alone had a flock of from forty to fifty thousand of the best English breeds, a number which has since greatly increased. The wool is said to be remarkably fine in quality. In various parts of the islands the cattle although now nominally belonging to some proprietor or lessee are nearly wild; and the skill shown by the Buenos-Ayrean Guachos in hunting them down and capturing them with the bolas is very remarkable; the Scottish shepherds, many of whom have settled in the islands of late years, are, however, rapidly becoming as expert as their less civilized predecessors. A. wild dog was common on both islands some years ago, but on the east island it is now nearly exterminated.

On the day of our arrival, Captain Thomson and I paid our respects to the governor, Colonel D'Arcy, and we found him greatly interested in our visit owing to a report which had reached Stanley that some seams of graphite and workable beds of coal had been found at Port Sussex on the other side of the island. Although from the little I had seen and read of the geology of the islands, and still more from the appearance of the specimens shown me by Colonel D'Arcy, I felt pretty well assured that the quest would be fruitless, to satisfy the Governor and the agent of the Falkland Islands Company I asked Mr. Moseley, who was glad of the opportunity of seeing more of the country, to ride across and ascertain the true state of affairs. His observations justified our previous opinion. The whole of the east island, and probably the greater part of the west island also, consists of sedimentary rocks of palæozoic age; in the low grounds, clay-slate and soft sandstone, and on the ridges hardened sandstone passing into the conspicuous white quartzites. The beds of so-called coal were simply very bituminous beds among the clay-slates, sometimes becoming a sort of culm, which might possibly answer to mix with coal and burn in a smithy fire, like the bituminous slates in the Bala series of Tyrone and Dumfriesshire, but which could never be worked with advantage. The graphite was only the blackest samples of the same material.

Mr. Moseley brought back a fine lot of fossils from the sandstone, the beds and their contents having very much the appearance of the ferruginous sand-stones of May Hill or Girvan. The species of *Orthis, Atrypa,* and *Spirifer* are different ; and as there are no graptolites in the schists it is probable that the whole series belongs to a somewhat later period, possibly the base of the Devonians. But if Mr. Moseley did not find coal, he brought home slung at his saddle-bow what was of much greater interest to us—the skull and a great part of the skeleton of a rare little whale belonging to the genus *Xiphius.*

The Falkland Islands consist of the older palæozoic rocks, lower Devonian or upper Silurian, slightly metamorphosed and a good deal crumpled and distorted. It is entirely contrary to our experience that coal of any value should be found in such beds. Galena may occur in the quartzites, but probably in no great quantity ; and there is no positive reason why gold may not be found, although the beds have scarcely the character of auriferous quartz.

On our second visit to the town our eyes were refreshed by the vision of a Bishop ; not a bishop blunt of speech and careless of externals, as so hard-working a missionary among the Fuegians and Patagonians might well afford to be, but a bishop gracious in manner and perfect in attire, who would have seemed more in harmony with his surroundings in the atmosphere of Windsor or St. James's. We had great pleasure in the society of Bishop Stirling during our stay at Stanley. Although he takes his title from the Falklands, his diocese is so large—extending round the whole of the southern coast of South America—that his visits to Stanley are somewhat rare ; and we owed the pleasure of making his acquaintance to an accident which had befallen his little missionary schooner, the repair of which he was superintending. He is a most active and zealous pastor, and greatly beloved by his scattered flock. A great part of his time is spent in Fuegia, where he has succeeded in establishing a half-civilized missionary station, and it was most interesting to hear him talk of his strange experiences among perhaps the most primitive race in the world. Walking over the breezy ' Camp ' of the Falklands with Dr. Stirling, one could not help thinking that his great influence in these remote regions might to some extent be referred to the almost exaggerated care with which he maintains the culture and refinement of a gentleman and the dignity of the ecclesiastical office.

Two vegetable productions of the Falklands, the ' balsam-bog ' and the ' tussock-grass,' have been objects of curiosity and interest ever since the first accounts of the islands reached us. In many places the low ground looks from a little distance as if it were thickly scattered over with large grey boulders, hemispherical or oval, three or four feet high, and three or four to six or eight feet across. To heighten the illusion, many of these blocks are covered with lichens, and bunches of grass grow in soil collected in crevices, just as they would in little rifts in rocks. These boulder-like masses are single plants of *Bolax glebaria*, an umbellifer which has the strange habit which we had already seen in the *Azorella* of Kerguelen Island, only greatly exaggerated. These lumps of balsam-bog are quite hard and nearly smooth, and only when looked at closely they are seen to be covered with small hexagonal markings like the calyces on a weathered piece of coral. These are the circlets of leaves and the leaf-buds terminating a multitude of stems, which have gone on growing with extreme slowness and multiplying dichotomously for an unknown length of time, possibly for centuries, ever since the plant started as a single shoot from a seed. The growth is so slow, and the condensation from constant branching is so great, that the block becomes nearly as hard as the boulder which it so much resembles, and it is difficult to cut a shaving from the surface with a sharp knife. Under the unfrequent condition of a warm day with the sun shining, a pleasant aromatic odour may be perceived where these plants abound, and a pale yellow gum exudes from the surface which turns brown in drying. The gum is astringent and slightly aromatic, and the shepherds use it dissolved in spirit as a balsam for wounds and sores. The flowers, which are very inconspicuous, are produced at the ends of the branches, and the characteristic cremocarps of the umbelliferæ may be seen scattered over the smooth surface of the ball in late summer.

Bolax is uneatable, and can apparently be applied to no particular use, and as it is widely distributed and abundant it is likely that it will long hold its

place as one of the curiosities of the Falklands ; such
is unfortunately not a reasonable anticipation for that
prince of grasses *Dactylis cæspitosa.* The tussock-
grass grows in dense tufts from six to ten feet high ;
the leaves and stems are most excellent fodder and
extremely attractive to cattle, but the lower portions
of the stems and the crowns of the roots have un-
luckily a sweet nutty flavour, which makes them
irresistible, and cattle and pigs and all creatures,
herbivorous and omnivorous, crop the tussocks to the
ground, when the rain getting into the crowns rots
the roots ; or if they have the means, they tear them
out bodily. The work of extermination has proceeded rapidly, and now the tussock-grass is confined
to patches in a narrow border round the shore, and
to some of the outlying islands. When we were
lying off Port Louis at the head of Berkeley Sound
there was a pretty little islet thickly covered with
a perfectly even crop of tussock-grass about eight
feet high, and so dense that it could be mown
with a scythe ; we sent a boat's crew for a supply
for the animals on board, by whom it was highly
appreciated.

The peat of the Falkland Islands is very different
in character from that of the north of Europe ;
cellular plants enter scarcely if at all into its composition, and it is formed almost entirely of the roots
and matted foliage and stems of *Empetrum rub-
rum,* a variety of the common 'crow-berry' of the
Scottish hills with red berries, called by the Falk-
landers the 'diddle-dee' berry ; of *Myrtus nummularia,*
a little creeping myrtle which also produces red
berries with a pleasant flavour and leaves which are

used as a substitute for tea ; of *Caltha appendiculata,* a dwarf species of the marsh-marigold ; and of some sedges and sedge-like plants such as *Astelia pumila, Gaimardia australis,* and *Rostkovia grandiflora.* The roots and stems of these, preserved almost unaltered, may be traced down several feet into the peat, but finally the whole structure becomes obliterated, and the whole is reduced to an amorphous carbonaceous mass. The general flora of the ' Camp ' is much like that of the low grounds of Fuegia and Patagonia, but one misses the pretty flowering shrubs, especially the *Pernettyas,* and the lovely *Philesia buxifolia ;* the Smilaceæ are however still well represented by the beautiful and delicately perfumed ' almond-flower ' of the settlers, *Callixene marginata.*

The weather while we were at the Falklands was generally cold and boisterous, and boat-work was consequently uncomfortable and frequently impracticable except in the shallow water within the harbour ; we had however two or three days' dredging in the pinnace, and made a pretty fair account of the submarine inhabitants of our immediate neighbourhood. *Macrocystis pyrifera,* the huge tangle of the Southern Seas, is very abundant in Stanley Harbour, anchored in about ten fathoms, the long fronds stretching for many yards along the surface and swaying to and fro with the tide. Adhering to the fronds of macrocystis there were great numbers of an elegant little cucumber-shaped sea-slug (*Cladodactyla crocea,* LESSON, sp.) from 80 to 100 mm. in length by 30 mm. in width at the widest part, and of a bright saffron-yellow colour. The mouth and excretory opening are terminal ; ten long, delicate, branched oral tentacles, more resembling in form and attitude those of *Ocnus* than those of the typical *Cucumariœ*, surround the mouth : the perisom is thin and semi-transparent, and the muscular bands, the radial vessels, and even the internal viscera can be plainly seen through it. The three arterior ambulacral vessels are approximated, and on these the tentacular feet are numerous and well developed, with a sucking-disk supported by a round cribriform calcareous plate, or more frequently by several wedge-shaped radiating plates arranged in the form of a rosette; and these three ambulacra form together, at all events in the female, a special ambulatory surface.

The two ambulacral vessels of the 'bivium' are also approximated along the back, and thus the two interambulacral spaces on the sides of the animal, between the external trivial ambulacra and the ambulacra of the bivium, are considerably wider than the other three; consequently, in a transverse section, the ambulacral vessels do not correspond with the angles of a regular pentagon, but with those of an irregular figure in which three angles are approximated beneath and two above. In the female the tentacular feet of the dorsal (bivial) ambulacra are very short; they are provided with sucking-disks, but the calcareous support of the suckers is very rudimentary, and the tubular processes are not apparently fitted for locomotion. In the males there is not so great a difference in character between the ambulacra of the trivium and those of the bivium; but the tentacles of the latter seem to be less fully developed in both sexes, and I have never happened to see an individual of either sex progressing upon, or adhering by, the water-feet of the dorsal canals.

In a very large proportion of the females which I examined, young were closely packed in two continuous fringes adhering to the water-feet of the dorsal ambulacra (Fig. 38). The young were in all the later stages of growth, and of all sizes from 5 up to 40 mm. in length; but all the young attached to one female appeared to be nearly of the same age and size. Some of the mothers with older families had a most grotesque appearance—their bodies entirely hidden by the couple of rows, of a dozen or so each, of yellow vesicles like ripe yellow plums ranged along their backs, each surmounted by its expanded crown of oral tentacles; in the figure the young are represented about half-grown. All the young I examined were miniatures of their parents; the only marked difference was that in the young the ambulacra of the bivium were quite rudimentary—they were externally represented only by bands of a somewhat darker orange than the rest of the surface, and by lines of low papillæ in the young of larger growth; the radial vessels could be well seen through the transparent body-wall; the young attached themselves by the tentacular feet of the trivial ambulacra, which are early and fully developed.

We were too late at the Falklands (January 23) to see the process of the attachment of the young in their nursery, even if we could have arranged to keep specimens alive under observation. There can be little doubt that, according to the analogy of the class, the eggs are impregnated either in the ovarial tube or immediately after their extrusion, that the first developmental stages are run through rapidly, and that the young are passed back from the ovarial

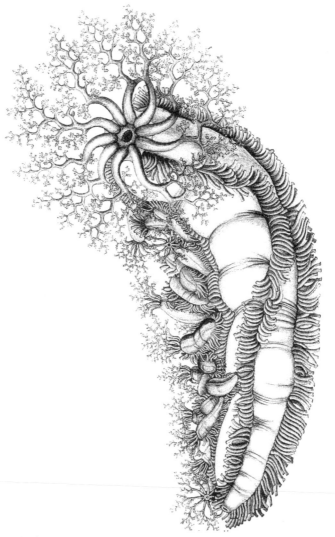

FIG. 38.—*Cladodactyla crocca*, LESSON. Stanley Harbour, Falkland Islands. Natural size.

pening, which is at the side of the mouth, along

the dorsal ambulacra, and arranged in their places by the automatic action of the ambulacral tentacles themselves.

The very remarkable mode of reproduction of certain members of all the recent classes of Echinodermata by the intervention of a free-swimming bilaterally symmetrical 'pseudembryo' developed directly from the 'morula,' from which the true young is subsequently produced by a process of internal budding or rearrangement, has long been well known through the labours of a host of observers headed and represented by the late illustrious Professor Johannes Müller, of Berlin.

At the same time it has all along been fully recognised that reproduction through the medium of a 'pseudembryo' is not the only method observed in the class; but that in several of the Echinoderm orders, while in a certain species a wonderfully perfect and independent bilateral locomotive zooid may be produced, in very nearly allied species the young Echinoderm may be developed immediately from the segmented yelk without the formation of a 'pseudembryo,' or at all events with no further indication of its presence than certain obscure temporary processes attached to the embryo, to which I have elsewhere (*Phil. Trans.* for 1865, p. 517) given the name of 'pseudembryonic appendages.'

This direct mode of development has been described in *Holothuria tremula* by MM. Koren and Danielssen, in *Synaptula vivipara* by Professor Oersted, in a 'viviparous sea-urchin' by Professor Grube, in *Echinaster* and in *Pteraster* by Professor Sars, in *Asteracanthion* by Professor Sars, Professor Agassiz,

Dr. Busch and myself, in *Ophiolepis squamata* by Professor Max Schultze, and in 'a viviparous ophiurid' by Professor Krohn. No less than four of these observations were made on the coast of Scandinavia. In temperate regions, where the economy of the Echinoderms has been under the eye of a greater number of observers, the development of the free-swimming larva appeared to be so entirely the rule that it is usually described as the normal habit of the class; while on the other hand, direct development seemed to be most exceptional. I was therefore greatly surprised to find that in the southern and subarctic seas a large proportion of the echinoderms of all orders, with the exception perhaps of the crinoids (with regard to which we have no observations), develop their young after a fashion which precludes the possibility, while it nullifies the object, of a pseudembryonic perambulator, and that in these high southern latitudes the formation of such a locomotive zooid is apparently the exception.

This modification of the reproductive process consists in all these cases, as it does likewise in those few instances in which direct development has already been described, of a device by which the young are reared within or upon the body of the parent, and are retained in a kind of commensal connection with her until they are sufficiently grown to fend for themselves. The receptacle, in cases where a special receptacle exists in which the young are reared, has been called a 'marsupium' (Sars), a term appropriately borrowed from the analogous arrangement in their neighbours the aplacental mammals of Australia. The young do not appear to have in any case an organic connection with the parent; the impregnated egg from the time of its reaching the 'morula' stage is entirely free; the embryos are indebted to the mother for protection, and for nutrition only indirectly through the mucus exuded from the surface of her perisom, and through the currents of freshly aerated water containing organic matter brought to them or driven over them by the action of her cilia.

Animals hatching their eggs in this way ought certainly to give the best possible opportunities for studying the early stages in the development of their young. Unfortunately, however, this is a kind of investigation which requires time and stillness and passable comfort; and such are not the usual conditions of a voyage in the Antarctic sea. Specimens have been carefully preserved with the young in all stages; and I hope that a careful examination of these may yield some further results.

Cladodactyla crocea is one of the forms in which there is no special marsupium formed; it is possible that the comparatively genial condition of the land-locked fiords and harbours of the Malvinas, and the additional shelter yielded by the imbricating fronds of *Macrocystis*, may render such exceptional provision unnecessary.

Five at least of these directly developing echinoderms representing five principal divisions of the sub-kingdom, were dredged at the Falklands, and several others were found earlier in the voyage in the sub antarctic regions of the Southern Sea. It will perhaps give a better idea of the diversity of means by which practically the same end is attained, if I give here a brief description of the principal modifications of the process which were exhibited.

To give a second example from the *Holothuroidea*, on the morning of the 7th of February, 1875, we dredged at a depth of 75 fathoms, at the entrance of Corinthian Harbour (*alias* 'Whisky Bay') in Heard Island (so far as I am aware the most desolate spot on God's earth), a number of specimens of a pretty

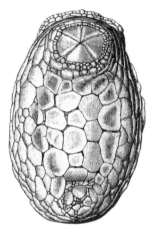

FIG. 39.—*Psolus ephippifer*, WYVILLE THOMSON. Corinthian Harbour, Heard Island. Three times the natural size.

little *Psolus*, which I shall here call for the sake of convenience *P. ephippifer*, although it may very possibly turn out to be a variety of the northern *P. operculatus*.

P. ephippifer (Figs. 39, 40) is a small species, about 40 millims. in length by 15 to 18 millims. in extreme width. In accordance with the characters of the genus, the ambulatory area is abruptly defined, and tentacular feet are absent on the upper surface of the body, which is covered with a thick leathery membrane in which calcareous scales of irregular form are imbedded. The oral and excretory openings are on the upper surface, a little behind the anterior border of the ambulatory tract, and a little in advance of the posterior extremity of the body respectively. A slightly elevated pyramid of five very accurately fitting calcareous valves closes over the oral aperture and the ring of oral tentacles ; and a less regular valvular arrangement covers the vent.

In the middle of the back in the female there is a well-defined saddle-like elevation formed of large tessellated plates somewhat irregular in form, with the surfaces smoothly granulated (Fig. 39). On removing one or two of the central plates we find that they are not, like the other plates of the perisom, imbedded partially or almost completely in the skin, but that they are raised up on a central column like a mushroom or a card-table, expanding above to the form of the exposed portion of the plate, contracting to a stem or neck, and then expanding again into an irregular foot, which is embedded in the soft tissue of the perisom ; the consequence of this arrangement is that when the plates are fitted together edge to edge, cloister-like spaces are left between their supporting columns. In these spaces the eggs are hatched, and the eggs or the young in their early stages are exposed by removing the plates (Fig. 40). At first, when there are only morules or very young embryos in the crypts, the marsupium is barely raised above the general surface of the perisom, and the plates of the marsupium fit accurately to one another ; but as the embryos increase in size, the marsupium projects more and more, and at length the joints between the plates begin to open (Fig. 39), and finally they open sufficiently to allow the escape of the young. The young in one marsupium seem to be all nearly of an age. In *P. ephippifer* the marsupium occupies the greater part of the dorsal surface, and its passages

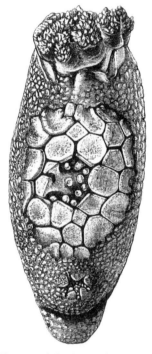

FIG. 40.—*Psolus ephippifer,* some of the plates of the marsupium removed. Three times the natural size.

run close up to the edge of the mouth, so that the eggs pass into them at once from the ovarial opening without exposure.

In the male there is, of course, no regular marsupium; but the plates are arranged in the middle of the back somewhat as they are in the female, except that they are not raised upon peduncles; so that it is not easy at once to distinguish a male from an infecund female.

Although we have taken species of *Psolus* sometimes in great abundance in various parts of the world, particularly in high latitudes, southern and northern, I have never observed this peculiar form of the reproductive process except on this one occasion.

On the 28th of January we dredged from the steam pinnace in about 10 fathoms water off Cape Pembroke, at the entrance of Stanley Harbour, a number of specimens of a pretty little regular sea-urchin *Goniocidaris canaliculata*, A. AGASSIZ.

The genus *Goniocidaris* (Desor) seems to differ from the genus *Cidaris* in little else than in having a very marked, naked, zigzag, vertical groove between the two rows of plates of each interambulacral area, and one somewhat less distinct between the ranges of ambulacral plates. It includes about half a dozen species, which appear to be mainly confined to the colder regions of the southern hemisphere, although two of the species extend as far to the northward as the East Indies and Natal.

This species (Fig. 41) has a general resemblance at a first glance to the small Mediterranean variety (*affinis*) of *Cidaris papillata*, but the radioles are thinner and much shorter, and differ wholly in their sculpture; the shell is even more depressed; the secondary tubercles are more distant; and a very regular series of short club-shaped rays seated on miliary granules are interposed in the rows between the spines of the second order. The ovarial openings are extremely minute, and are placed close to the outer edge of the

ovarial plates. The upper part of the test is quite
flat, the flat space including not only the ovarial
plates and the plates of the periproct, but the first
pair, at least, of the plates of each interambulacral
area. Articulated to the primary tubercles of these
latter are two circles of radioles, the inner more
slender and shorter, the outer stouter and longer,

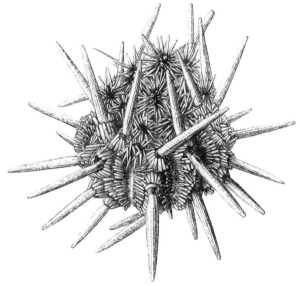

Fig. 41.—*Goniocidaris canaliculata,* A. Agassiz. Stanley Harbour. Twice the
natural size.

but both series much larger than radioles usually
are in that position on the test.

These special spines are cylindrical, and nearly
smooth, and they lean over towards the anal opening, and form an open tent for the protection of
the young, as in *Cidaris nutrix,* a species pre-
sently to be described, but at the opposite pole of the
body. In this species the eggs are extruded directly
into the marsupium ; and I imagine, from the very

small size of the ovarial openings, that when they
enter it, they are very minute, and probably unim-
pregnated.　In the examples which we dredged at the
Falkland Islands, the young were, in almost every
case, nearly ready to leave the marsupium ; we were
too late in the season to see the earlier stages.
young in the same marsupium are nearly all of an
age, some somewhat more advanced than others.
The diameter of the test is from 1 to 1·5 millim.,
and the height about ·8 millim. ; the length of the
primary spines is, in the most backward of the brood,
·5 millim., while in the most advanced it equals the
diameter of the test.　The perisom, in which the
cribriform rudiments of the plates of the corona and
the young spines are being developed, is loaded with
dark purple pigment, which makes it difficult to
observe the growth of the calcareous elements. About
thirty primary spines arise on the surface of the
corona almost simultaneously in ten rows of three
each : they first make their appearance as small
papillæ covered with a densely pigmented ciliated
membrane ; and when they have once begun to
lengthen, they run out very rapidly until they bear to
the young nearly the same proportions which the
full-grown spines bear to the mature corona.　Very
shortly some of the secondary spines, at first nearly
as large as the sprouting primary spines, make their
appearance in the interstices between these ; and a
crowd of very small spines rise on the nascent scales
of the peristome.　Successively five or six pedicellariæ
are developed towards the outer edge of the apical
area, which at this stage is disproportionately large ;
the pedicellariæ commence as purple papillæ, which

are at first undistinguishable from young primary
spines ; the first set look enormously large in proportion to the other appendages of the perisom. Almost
simultaneously with the first appearance of the
primary spines, ten tentacular feet, apparently the
first pairs on each ambulacrum of the corona, just
beyond the edge of the peristome, come into play ;
they are very delicate and extremely extensile, with
well-defined sucking-disks ; and with these the young
cling to and move over the spines of the mother, and
cling to the sides of the glass vessel, if they are dislodged from the marsupium. This species seems to
acquire its full size during a single season. We
dredged it at the close of the breeding season, and we
took no specimens intermediate in size between the
adult and the young.

Among the marine animals which we dredged from
the steam-pinnace on the 19th of January, 1874, at
depths of from 50 to 70 fathoms in Balfour Bay (a fine
recess of one of the many channels which separate the
forelands and islands at the head of Royal Sound,
Kerguelen Island), there were several examples of
a small *Cidaris*, which I will name provisionally
C. nutrix (Fig. 42).

This species resembles *C. papillata* in the general
form and arrangement of the plates of the corona, in
the form and arrangement of the primary tubercles
of the interambulacral areas and of the secondary
tubercles over the general surface of the test, in the
form of the plates of the apical disk and of the imbricated calcareous scales of the peristome, in the form,
sculpture, and proportionate length of the primary
spines, and in the form of the different elements of the

jaw-pyramid and in that of the teeth ; but the test is
more depressed, the secondary spines which articulate
to the ambulacral plates and cover the pore-areas are
longer and more cylindrical, not so much flattened as

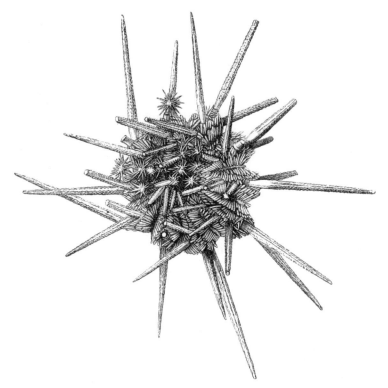

Fig. 42.—*Cidaris nutrix*, Wyville Thomson.   Balfour Bay, Kerguelen Island.
Natural size.

they are in *C. papillata ;* the large tulip-like pedicellariæ and the long thin tridactyle pedicellariæ mixed
with the secondary spines in the northern species are
wanting, or in very small number ; and the minute
pedicellariæ of the peristome are much fewer. The
ovaries, which in *C. papillata* have the walls loaded

with large expanded calcareous plates, contain only
a few small branched spicules; and the calcareous
bodies in the wall of the intestine are small and
distant.   The perforations in the ovarial plates in
the female are somewhat larger than in *C. papillata;*
and the ripe ova in the ovary appear to be considerably larger.

The eggs, after escaping from the ovary, are passed
along on the surface of the test towards the mouth;
and the smaller slightly spathulate primary spines
which are articulated to about the first three rows of
tubercles round the peristome, are bent inwards over
the mouth, so as to form a kind of open tent, in which
the young are developed directly from the egg without undergoing any metamorphosis, until they have
attained a diameter of about 2·5 mms.; they are then
entirely covered with plates, and are provided with
spines exceeding in length the diameter of the test.
Even before they have attained this size and development, the more mature or more active of a brood may
be seen straying away beyond the limits of the
'nursery,' and creeping with the aid of their first
few pairs of tentacular feet out upon the long
spines of their mother; I have frequently watched
them return again after a short ramble into the
'marsupium.'

I am not aware that a free pseudembryo, or
'pluteus,' has been observed in any species of the
restricted family Cidaridæ; but I feel very certain
that *Cidaris papillata* in the northern hemisphere,
except possibly in the extreme north, has no marsupial
arrangement such as we find in the Kerguelen *Cidaris.*
There have passed through my hands during the last

few years hundreds of specimens of the normal northern form, of the Mediterranean varieties *C. hystrix* and *C. affinis* (*stokesii*), and of the American *C. abyssicola*, from widespread localities and of all ages ; and I have never found the young except singly, and never in any way specially associated with breeding individuals.

In Stanley Harbour we dredged many specimens of an irregular urchin, much resembling in general

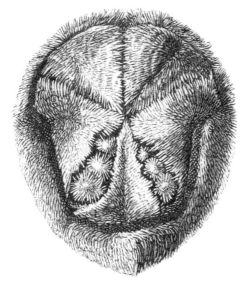

FIG. 43.—*Hemiaster philippii*, GRAY. Accessible Bay, Kerguelen Island. Twice the natural size.

appearance *Brisopsis lyrifera*, the common 'fiddle urchin' of the boreal province of the British Seas, and probably to be referred to *Hemiaster philippii*, GRAY.

These urchins were not breeding when we were at the Falklands, but on the 9th of January, 1874, we

dredged from the pinnace in shallow water, varying
from 20 to 50 fathoms, with a muddy bottom, in
Accessible Bay, Kerguelen Island, innumerable
samples of apparently the same species.

The test of a full-sized example (Fig. 43) is about
45 mms. in length and 40 mms. in width; the height
of the shell in the female is 25 mms., in the male it
is considerably less. The apex is nearly in the centre
of the dorsal surface; the genital openings are three
in number, in the female very large; the bilabiate
mouth is placed well forward on the ventral aspect;

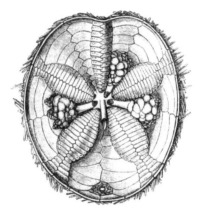

FIG. 44.—*Hemiaster philippii.* The apical portion of the test of the female seen from
within. Slightly enlarged.

and the excretory opening is posterior and supra-
marginal. The odd anterior ambulacrum is shallow,
and the tube-feet which are projected from it are large
and capitate. The anterior paired ambulacra are
somewhat longer than the posterior. The whole of
the surface of the test is covered with a close pile of
small spines of a dark green colour; those fringing
the ambulacral grooves are long and slightly curved,

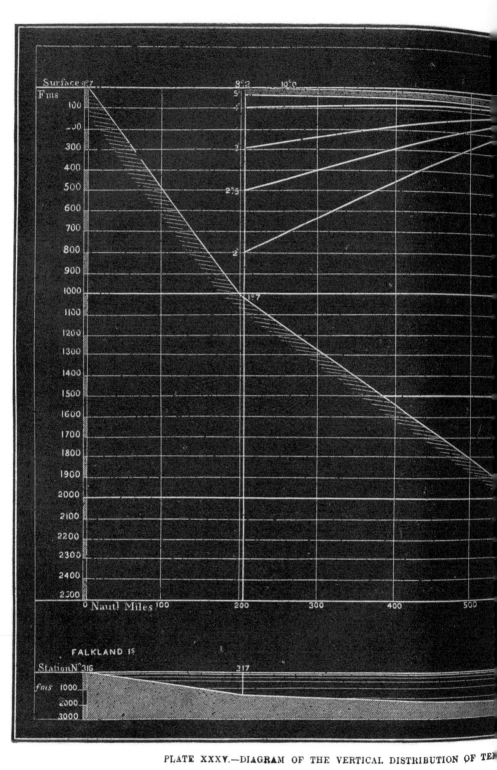

PLATE XXXV.—DIAGRAM OF THE VERTICAL DISTRIBUTION OF TE

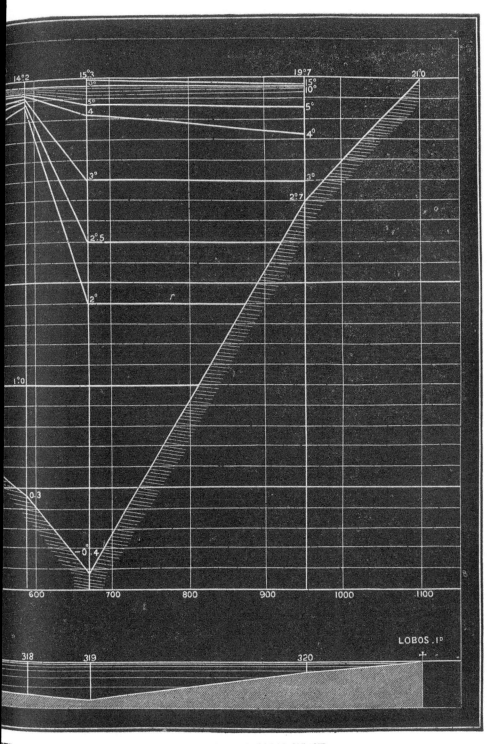

and they bend and interdigitate so accurately over the
ambulacra that one might easily overlook the grooves
at a first glance. The peripetalous fasciole is somewhat irregular ; but in those examples in which it is
best defined it forms a wide arch, extending backwards
on each side a little beyond the lateral ambulacra of
the trivium, and then, contracting a little, forms a
rudely rectangular figure round the bivium. The
paired ambulacral grooves in the male are shallow,
not much deeper than the anterior ambulacrum (Fig.
45) ; in the female the pore-plates of the paired

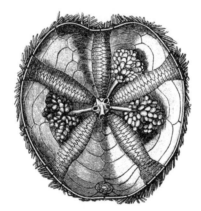

Fig. 45.—*Hemiaster philippii.* The apical portion of the test of the male seen from
within. Slightly enlarged.

ambulacra are greatly expanded and lengthened, and
thinned out and depressed so as to form four deep,
thin-walled, oval cups sinking into and encroaching
upon the cavity of the test and forming very efficient
protective 'marsupia' (Fig. 44). The ovarial openings are, of course, opposite the interradial areas ;
but the spines are so arranged that a kind of covered
passage leads from the opening into the marsupium ;

and along this passage the eggs, which are remarkably large, upwards of a millimetre in diameter when
they leave the ovary, are passed, and are arranged
very regularly in rows on the floor of the pouch, each
egg being kept in its place by two or three short
spines which bend over it (Fig. 46).

Among the very many examples of this *Hemiaster*

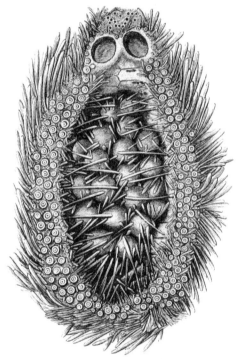

Fig. 46.—*Hemiaster philippii.* The arrangement of the eggs in one of the marsupial
recesses. Five times the natural size.

which we dredged in Accessible Bay, and afterwards
in Cascade Harbour, Kerguelen, there were young in
all stages in the breeding-pouches; and although from
the large size and the opacity of the egg and embryo it
is not a very favourable species for observation, had

other conditions been favourable, we had all the material
for working out the earlier stages in the development of
the young very fully. The eggs, on being first placed
in the pouches, are spherical granular masses of a deep
orange colour, enclosed within a pliable vitelline
membrane, which they entirely fill. They become
rapidly paler in colour by the development of the
blastoderm; they then increase in size probably by
the imbibition of water into the *gastrula* cavity; and
a whitish spot with a slightly raised border indicates
an opening which, I have no reason to doubt, is the
permanent mouth; but of this I cannot be absolutely
certain. The surface now assumes a translucent
appearance, and becomes deeply tinged with dark
purple and greenish pigment; and almost immediately, without any definite intermediate steps, the
outer wall is filled with calcified tissue, it becomes
covered with fine spines and pedicellariæ, a row of
tentacular feet come into action round the mouth, the
vent appears at the posterior extremity of the body,
and the young assumes nearly the form of the adult.
These later changes take place very quickly; but they
are accompanied by the production of so much heavy
purple and dark green pigment that it is difficult to
follow them. The viscera are produced at the expense
of the abundant yelk; and the animals at once take
a great start in size by the imbibition of water into
the previsceral cavity. The young urchins jostle one
another on the floor of the breeding-pouch, those
below pushing the others up until the upper set are
forced out between the rows of fringing spines of the
pouch; but even before leaving the marsupium, on
carefully opening the shell of the young, the intestine

may be seen already full of dark sand, following much the same course which it follows in the adult. The size of the test of the young on leaving the marsupium is about 2·5 mms. in length by 2 mms. in width.

We took along with the last species in Stanley Harbour several specimens of a large species of *Asteracantion* which formed a marsupium after the manner so well described by Sars in *Echinaster sarsii*, MULLER, by drawing its arms inwards and forwards, and forming a brood-chamber over the mouth. In some samples of this species the young were so far advanced that when the mother was placed in a jar they crept out of the nursery and wandered over the glass wall of their prison; this brood had entirely lost the ' pseudembryonic appendages,' but in their younger condition these are very apparent, though scarcely so well developed as in the young of *A. violaceus* on our own coast.

On the 27th of January, 1874, at Station 149, off Cape Maclear on the south-east coast of Kerguelen Island, we dredged a handsome starfish allied to *Luidia* or *Archaster*, which has since been described by Mr. Edgar Smith from specimens brought home by the Rev. Mr. Eaton, under the name of *Leptychaster kerguelenensis* (Fig. 47).

A well-grown example is from 100 to 120 mm. in diameter from tip to tip of the arms; the length of the arm is about three times its width near the base, and three times the diameter of the disk. The marginal plates are long and narrow, running up with a slight curve outwards from the edge of the ambulacral groove until they meet the border of the

dorsal perisom above; they are closely set with short
blunt spines, which become gradually a little longer
towards the radial groove; and at the edge of the

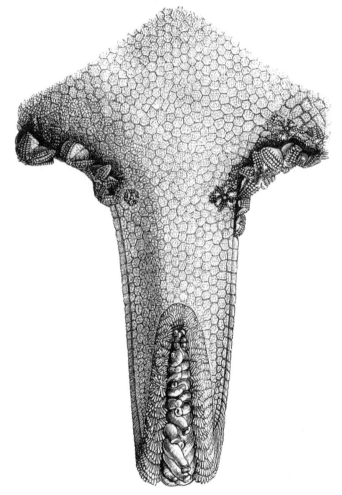

Fig. 47.—*Leptychaster kerguelenensis*, E. SMITH.　Off Cape Maclear, Kerguelen Island.
Twice the natural size.

groove each plate bears a tuft of about six rather long
spines: these tufts in combination form a scalloped

fringe spreading inwards on each side over the groove. The dorsal surface of the body is covered with a tesselated pavement composed of capitate paxilli. The heads of the paxilli in close apposition combine to form a mosaic with rudely hexagonal facets; and as they are raised upon somewhat slender shafts, whose bases, like the plinths of columns, rest upon the soft perisom, arcade-like spaces are left between the skin and the upper calcareous pavement. The eggs pass into these spaces from the ovarial openings: on bending the perisom and separating the facets, they may be seen in numbers among the shafts of the paxilli. There is a continual discharge of ova into the passages, so that eggs and young in different stages of development occupy the spaces at one time. The young do not escape until at least six ambulacral suckers are formed on each arm; they may then be seen pushing their way out by forcing the paxilli to the side, and squeezing through the chink between them. While it is extricating itself the oral surface of the young is always above: and the centre of the star with the mouth is usually the part which first protrudes; then the arms disengage themselves one after another, many of the brood remaining for a time with one or two arms free and the others still under the paxilli. When the young have become disengaged, they remain for a considerable time attached to the parent by the centre of the dorsal surface. I could never satisfy myself by what means this is effected; the attachment is very slight, and they are removed by the least touch. In this attached stage until they entirely free themselves, which they do when the number of tentacular feet on

CHAP. IV.] *THE VOYAGE HOME.* 237

each arm has reached about twenty, they cluster in the re-entering angles between the arms of the mother, spreading a little way along the arms and on the dorsal surface of the disk; the young escape from the marsupium chiefly in the neighbourhood of the angles between the rays. The madreporiform tubercle is visible in the young near the margin of the disk between two of the arms; but in the mature starfish it is completely hidden by the paxilli, and no doubt it opens into the space beneath them.

We took *Leptychaster* in the act of bringing forth young on that one occasion only; and the weather was so boisterous at the time that it was impossible to trace the early stages in the development of the embryo. It is evident that the process generally resembles that described by Professor Sars in *Pteraster militaris;* and it is quite possible that, while there is certainly not the least approach to the formation of a locomotive bipinnaria, as in that species, some provisional organs may exist at an early period.

In 'The Depths of the Sea' (p. 120) I noticed and figured a singular little starfish from a depth of 500 fathoms off the north of Scotland under the name of *Hymenaster pellucidus.* This form was at that time the type of a new genus; but the researches of the last three years have shown that, with the exception perhaps of *Archaster, Hymenaster* is the most widely distributed genus of Asterids in deep water. It is met with (sparingly, it is true, only one or two specimens being usually taken at once in the trawl) in all parts of the great ocean; and it ranges in depth from 400 to about 2,500 fathoms.

On the 7th of March, 1874, we dredged an extremely handsome new form, to which I shall give provisionally the name of *H. nobilis,* in lat. 50° 1′ S., long. 123° 4′ E., 1,099 miles south-west of Cape Otway, Australia, at a depth of 1,800 fathoms, with a bottom of 'globigerina-ooze,' and a bottom-temperature of 0°·3 C.

Hymenaster nobilis (Fig. 48) is 300 mm. in diameter from tip to tip of the rays ; the arms are 55 mm. wide ; and, as in *H. pellucidus,* a row of spines fringing the ambulacral grooves are greatly lengthened and webbed, and the web running along the side of one arm meets and unites with the web of the adjacent arm, so that the angles between the arms are entirely filled up by a fleshy lamina stretched over and supported by spines, the body thus becoming a regular pentagon. The upper surface of the body, the disk, and the arms,—all the surface except the smooth membrane between the arms, is covered with fascicles of four to six diverging spines. These spines are about 3 mm. in height ; and they support and stretch out a tolerably strong membrane clear above the surface of the perisom, like the canvas of a marquee, leaving an open space beneath it. A close approach to this arrangement occurs also in *Pteraster.*

At the apical pole the upper free membrane runs up to and ends at a large aperture, 15 mm. in diameter, surrounded by a ring of five very beautifully formed valves. These valves do not essentially differ from the ordinary radiating supports of the marsupial tent ; a stout calcareous rod arises from the end of the double chain of ossicles which form the floor of the ambulacral groove. From the outer aspect of this support three or four spines diverge in the ordinary way under the tent-cover; but from its inner aspect six or eight slender spines rise in one plane with a special membrane stretched between them. When the valves are raised and the pentagonal chamber beneath them open, these spines separate from one another, and, like the ribs of a fan, spread out the membrane in a crescentic form (Fig. 48); and when the valves close, the spines approximate and are drawn downwards, the five valves forming together a very regular, low, five-sided pyramid (Fig. 49). Looking down into the chamber when the valves are raised, the vent is seen on a small projecting papilla in the centre of the floor; and between the supporting ossicles of the valves, five dark open arches lead into the spaces opposite the re-entering angles of the arms, which receive the ducts of the ovaries. In the particular specimen to which I have referred, which is considerably the largest of the genus which we have yet met with, there were one or two eggs in the pouch, but they were apparently abortive. It seemed that the brood had been lately discharged; for some oval depressions still remained on the floor of the central chamber, in which the eggs or the young had evidently been lodged. I have on three occasions in species of the genus *Hymenaster* found the eggs beneath the membrane in the angles of the arms, and, in a more advanced stage, congregated in the central tent, but never under circumstances such that I could keep and examine them; exposed or loosely covered eggs or embryos, or any soft and pulpy organs or appendages, are always

in a half disintegrated state when they are brought
up from such great depths, if they have not been
entirely washed away.

As I have already said, *Hymenaster* is closely allied

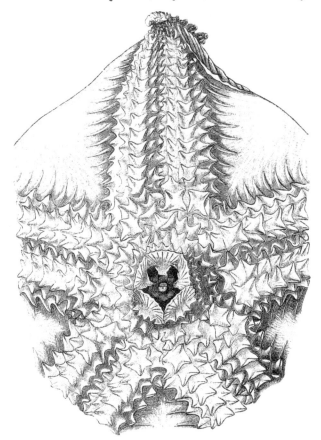

FIG. 48.—*Hymenaster nobilis*, WYVILLE THOMSON. Southern Sea. Half the natural size.

to *Pteraster :* the arrangements of the marsupium
are nearly the same in both; and it is highly probable
that, in *Hymenaster,* as in *P. militaris,* a provisional
alimentary tract may be developed in the early
stages of the embryo.

There are several fine species of *Hymenaster* within reach of British naturalists in the deep water at the entrance of the Channel and off Cape Clear; but I fear there will be great difficulty in determining this point unless the genus turn up somewhere in shallower soundings where specimens can be taken alive.

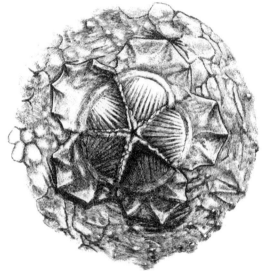

FIG. 49.—*Hymenaster nobilis.* The marsupial tent with the valves closed. Twice the natural size.

In Stanley Harbour, on the roots of Macrocystis, and also brought up free by the dredge, there were numerous examples of an Ophiurid which appears to correspond with *Ophiacantha vivipara*, LJUNGMAN; we had previously got either the same or a very closely allied form in great abundance in the Fjords of Kerguelen. The Kerguelen variety has been noticed by Mr. Edgar Smith, under the name of *Ophioglypha hexactis*, and I have called it, provisionally, in a paper in the *Proceedings* of the Linnæan Society, *Ophiocoma didelphis,* from its opossum-like habit of carrying its young upon its back.   I do not think that it can properly be relegated to any genus at present defined,

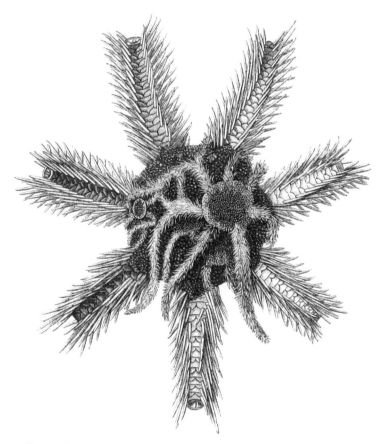

FIG. 50.—*Ophiocoma ? vivipara* (LJUNGMAN, Sp.).   Twice the natural size.   No. 149.

but it will doubtless fall into its place when the Ophiurids shall have been revised.

The disk is about 20 mm. in diameter ; and the arms are four times the diameter of the disk in length.   The disk is uniformly coarsely granulated ; the arm-shields, which are well defined through the membrane, are rounded in form and roughly granulated like the remainder of the disk.   The character which at once distinguishes this species from all the others of the genus is, that the normal number of the arms is six or seven instead of five, which is almost universal in the class.   The number of arms is subject to certain variation.   I have seen from six to nine, but never fewer than six.   The arm-spines are numerous and long.   The general colour of the disk and arms is a dull greenish brown.

A large proportion of the mature females, if not all of them, had a group of from three to ten or twelve young ones clinging to the upper surface of the disk by their arms : the largest of these were about a quarter the size of their mother ; and they graduated down in size until the smallest had a diameter of less than 1·5 mm. across the disk.   The largest and oldest of the progeny were always uppermost, furthest from the disk, the series decreasing in size downwards, and the supply evidently coming from the genital clefts beneath.   In several specimens which I examined, although by no means in all, there were groups of eggs and of young in still earlier stages, free in the body-cavity in the interbrachial spaces.

It thus seems that in this case the true ' marsupium ' is a portion of the body-cavity, and that the protection afforded by it is supplemented by the attachment of the young to the surface of the disk, maintained for some time after their extrusion or escape.

The process of propagation in *Ophiocoma vivipara*

differs from most of the other cases described, in the
eggs being successively hatched, and the young being
found consequently in a regularly graduated series of
stages of growth. Although I had not an opportunity
of working the matter out with the care and com-
pleteness I could have wished, I feel satisfied, from
the examination of several of the young at a very
early period, that in this case no provisional mouth
and no pseudembryonic appendages whatever are
formed, and that the primary aperture of the *gastrula*
remains as the common mouth and excretory opening
of the mature form. From the appearance of the
ovaries and of the broods of young, I should think it
probable that this species gives off young in a continuous series for a considerable length of time,
probably for some months.

I have selected these illustrations of the development of the young of echinoderms from the egg
without the intervention of a locomotive 'pseudembryo' from a much larger number. As I have
already said, I cannot, on account of the unfavourable
conditions for carrying on such investigations under
which the majority of the species were procured, say
with certainty that no trace of pseudembryonic
appendages or provisional organs exist in any of
these instances, but I feel satisfied that none such
occurs in *Psolus ephippifer*, in *Hemiaster philippii* or
in *Ophiacoma vivipara*. Neither am I in a position to
state that in these southern latitudes direct development is universal in the sub-kingdom. I believe
indeed that it is not so; for species of the genera
Echinus, Strongylocentrotus, and *Amblypneustes* run
far south, and a marsupial arrangement seems

improbable in any of these. It is, however, a sig-
nificant fact that, while in warm and temperate seas
'plutei' and 'bipinnariæ' are constantly taken in
the surface-net, in the southern sea they are almost
entirely absent.

Amid all their general tameness the Falkland
Islands boast one natural phenomenon which is
certainly very exceptional, and at the same time very
effective.

In the East Island most of the valleys are occupied
by pale grey glistening masses, from a few hundred
yards to a mile or two in width, which look at a
distance much like glaciers descending apparently
from the adjacent ridges, and gradually increasing in
volume, fed by tributary streams, until they reach the
sea. Examined a little more closely, these are found
to be vast accumulations of blocks of quartzite, irregular in form, but having a tendency to a rude diamond
shape, from two to eight or ten or twenty feet long
and perhaps half as much in width, and of a thickness
corresponding with that of the quartzite bands in the
ridges above. The blocks are angular like the fragments in a breccia, and they rest irregularly one upon
the other, supported in all positions by the angles
and edges of those beneath.

They are not weathered to any extent, though the
edges and points are in most cases slightly rounded;
and the surface, also perceptibly worn but only by the
action of the atmosphere, is smooth and polished;
and a very thin, extremely hard, white lichen which
spreads over nearly the whole of them gives the effect
of their being covered with a thin layer of ice.

Far down below, under the stones, one can hear the

stream of water gurgling which occupies the axis
of the valley ; and here and there, where a space
between the blocks is unusually large and clear, a
quivering reflection is sent back from a stray
sunbeam.

At the mouth of the valley the section of the
' stone river' exposed by the sea is like that of a stone
drain on a huge scale, the stream running in a
channel arched over by loose stone blocks, or finding
its way through the spaces among them.   There is
scarcely any higher vegetation on the ' stone-run ;'
the surface of every block is slippery and clean,
except where here and there a little peaty soil
has lodged in a cranny, and you find a few trailing spikes of *Nassauvia serpens,* or a few heads of
the graceful drooping chrysanthemum-like *Chabræa
suaveolens.*

These ' stone-rivers' are looked upon with great
wonder by the shifting population of the Falklands,
and they are shown to visitors with many strange
speculations as to their mode of formation.   Their
origin seems however to be obvious and simple
enough, and on that account their study is all the
more instructive, for they form an extreme case of a
phenomenon which is of wide occurrence, and whose
consequences are, I believe, very much underrated.

There can be no doubt that the blocks of quartzite
in the valleys are derived from the bands of quartzite
in the ridges above, for they correspond with them
in every respect; the difficulty is to account for
their flowing down the valley, for the slope from
the ridge to the valley is often not more than six
to eight degrees, and the slope of the valley itself

only two or three, in either case much too low to cause blocks of that form either to slide or to roll down.

The process appears to be this. The beds of quartzite are of very different hardness; some are soft, passing into a crumbling sandstone, while others are so hard as to yield but little to ordinary weathering. The softer bands are worn away in process of time, and the compact quartzites are left as long projecting ridges along the crests and flanks of the hill-ranges. When the process of the disintegration of the softer beds has gone on for some time the support of their adjacent beds is taken away from the denuded quartzites, and they give way in the direction of the joints, and the fragments fall over upon the gentle slopes of the hillside. The vegetation soon covers the fallen fragments, and usually near the sloping outcrops of the hard quartz, a slight inequality only in the surface of the turf indicates that the loose blocks are embedded beneath it. Once embedded in the vegetable soil, a number of causes tend to make the whole soil-cap, heavy blocks included, creep down even the least slope. I will only mention one or two of these. There is constant expansion and contraction of the spongy vegetable mass going on, as it is saturated with water or comparatively dry, and while with the expansion the blocks slip infinitesimally down, the subsequent contraction cannot pull them up against their weight; the rain water trickling down the slope is removing every movable particle from before them; the vegetable matter on which they are immediately resting is undergoing a perpetual process of interstitial decay and removal. In this way the blocks are gradually borne down the slope in the soil-cap and piled in the valley below. The only other question is how the soil is afterwards removed and the blocks left bare. This, I have no doubt, is effected by the stream in the valley altering its course from time to time, and washing away the soil from beneath.

This is a process which in some of the great ' stone-rivers ' in the Falkland Islands must have taken an enormous time. I fear that the extreme glacialists will see in it a danger to this universal application of their beloved theory to all cases of scratching and grooving. I have known too much of the action of ice to have the slightest doubt of its power; but I say that ice had no hand whatever in the production of these grand 'moraines' in the Falkland Islands.

In the West Highlands of Scotland, and in many other parts of the world, I have often noticed that when a hill of such a rock as clay-slate comes down with a gentle slope, the outcrop of the vertical or highly inclined slates covered with a thick layer of vegetable soil or drift containing embedded blocks and boulders derived from higher levels, the slates are frequently first slightly bent downwards, then abruptly curved and broken, and frequently the lines of the fragments of the fractured beds of slate can be traced for a yard or two in the soil-cap gradually becoming parallel with its surface, and passing down in the direction of its line of descent. These movements are probably extremely slow; I well remember many years ago observing a case somewhere in the West of Scotland, where a stream had exposed a fine section of the soil-cap with the lines of broken down

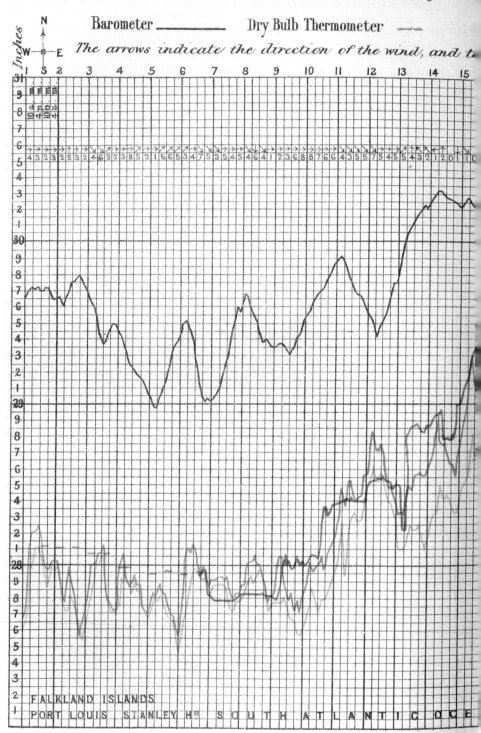

PLATE XXXVI. *Meteorological C*

Barometer ——————— Dry Bulb Thermometer ——

The arrows indicate the direction of the wind, and t

FALKLAND ISLANDS

PORT LOUIS   STANLEY HR   SOUTH ATLANTIC OCE

ervations for the month of *February*, 1876.

et Bulb Thermometer ——   Temperature of Sea Surface ——

numbers beneath its force according to Beaufort's scale

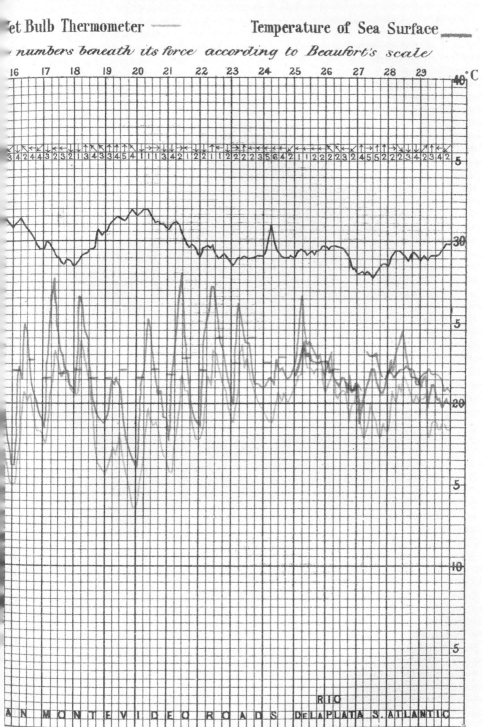

and crushed slate-beds carried far down the slope. The whole effect was so graphically one of vigorous and irresistible movement, that I examined carefully some cottages and old trees in hope of finding some evidence of twisting or other irregular dislocations, but there appeared to be none such ; the movement, if it were sufficiently rapid to make a sign during the life of a cottage or a tree, evidently pervaded the whole mass uniformly.

It seems to me almost self-evident that wherever there is a slope, be it ever so gentle, the soil-cap must be in motion, be the motion ever so slow ; and that it is dragging over the surface of the rock beneath the blocks and boulders which may be embedded in it ; and frequently piling these in moraine-like masses, where the progress of the earth-glacier is particularly arrested as at the contracted mouth of a valley, where the water percolating through among them in time removes the intervening soil. As the avalanche is the catastrophe of ice-movement, so the land-slip is the catastrophe of the movement of the soil-cap.

As I have already said, I should be the last to undervalue the action of ice, or to doubt the abundant evidences of glacial action ; but of this I feel convinced, that too little attention has been hitherto given to this parallel series of phenomena, which in many cases it will be found very difficult to discriminate ; and that these phenomena must be carefully distinguished and discriminated before we can fully accept the grooving of rocks and the accumulation of moraines as complete evidence of a former existence of glacial conditions.

On the first of February we went round to the head
of Berkeley Sound, and saw the old station of St.
Louis now nearly deserted, some shepherds in the
employment of the Falkland Islands Company having
occupied the old Government buildings. We returned
to Stanley on the 4th, and on the 6th we sailed for
Monte Video, bidding a final farewell to the Falklands, which I am sure we shall always remember
with pleasure, if not on their own account, on that of
the kindness and hospitality which we met with
during our stay.

On the 8th we sounded about 200 miles to the
north-east of Stanley, in a depth of 1,035 fathoms.
The sounding machine brought up no sample of the
bottom, but a tow-net attached to the dredge-rope
at the weight contained a little gravel and one or
two small organisms. The bottom-temperature was
$1°\cdot7$ C. The trawl was lowered, but it was unfortunately carried away; after the weights, which were
300 fathoms in advance of the trawl, had been brought
on board. The rope was much chafed, as if it had been
dragged against sharp rocks. The following day was
fine, with light, uncertain winds; on the 10th it was
blowing half a gale, and the sea was running too high
for sounding operations. On the 11th the weather
was fine, the wind becoming more moderate towards
noon ; at 10 A.M. we sounded and put down the trawl
in 2,040 fathoms, with a bottom of bluish mud containing many *Globigerinæ*, and a bottom-temperature
of $+ 0°\cdot3$ C. The position of the sounding was
Lat. 42° 32′ S., Long. 56° 27′ W., about 200 miles to
the eastward of Valdes Peninsula. Temperature
soundings were taken at this station down to 1,500

fathoms. This sounding gives a singularly rapid fall from 14°·2 C. on the surface to 2° C. at 125 fathoms; the edge of the Antarctic indraught appeared to be pushed up against the American shore by the western border of the southern branch of the reflux of the equatorial current, just as the Labrador current is banked up by its northern branch; the result being no doubt increased in both cases by the flinging up of the polar water against the western land-barrier on account of its low initial velocity. The trawl yielded only one or two fishes, some medusæ and a caridid shrimp, so that there was no actual evidence of its having reached the bottom.

On the 12th we sounded in 2,425 fathoms, and took a series of temperatures. The upper temperatures were decidedly higher than they were the day before, 5° C. occurring at 125 fathoms, 2°·5 C. at 700, and 2° C. at 1,100 fathoms. The position of the sounding was Lat. 41° 45′ S., Long. 54° 46′ W.; it was nearly double the distance of the previous sounding from the 100-fathom line, which very nearly corresponds with a submarine cliff of great height. The bottom-temperature was — 0°·4 C. On the 14th we sounded in 600 fathoms on the plateau extending from the South American coast, opposite the estuary of the River Plate, 144 miles from Labos Island. We took a set of temperatures to the bottom, and found the gradation, so far as it went, very much the same as on the 12th. The bottom-temperature was 2°·7 C. On this occasion the trawl was most successful, and gave us a good idea of the fauna of moderate depths along the coast; probably

not fewer than sixty species of different groups were
recovered, including a very handsome Pennatulid
between two and three feet in height, some deep-sea
corals of very special interest, and some fine
echinoderms and sponges.   On the 15th we anchored
in Monte Video Roads.

We left the anchorage of Monte Video at daybreak
on the 25th of February, and after swinging ship
for errors of the compasses we proceeded down the
estuary.   In the afternoon the trawl was put over
in thirteen fathoms to get an idea of the fauna of
the brackish water.   The species procured were
comparatively few, but among them was a plentiful
supply of an interesting alcyonarian of the genus
Renilla, which, although well known, had not been
met with by us before.   On the two following days
we crossed the shallow-water plateau, and on the
28th we sounded and trawled in 1,900 fathoms, over
the ledge.   The serial temperature-sounding gave
a bottom-temperature of $0°·0$ C.; at 1,725 the temperature was $1°·0$ C., at 600 fathoms $3°·0$ C., and at
50 fathoms $20°·0$ C.   The trawl was not very successful, but it brought up a few things of some
interest, among them an example of a small sea-
urchin (*Aceste bellidifera*), of which we had previously
taken single specimens at widely different stations,
off the coast of Nova Scotia, near Gomera Island,
near New Zealand, and near Japan.   The bottom
was chiefly river-mud, with very little carbonate
of lime.

On the following day we sounded in 2,800 fathoms
and again lowered the trawl.   The bottom was a
greyish mud with little or no carbonate of lime,

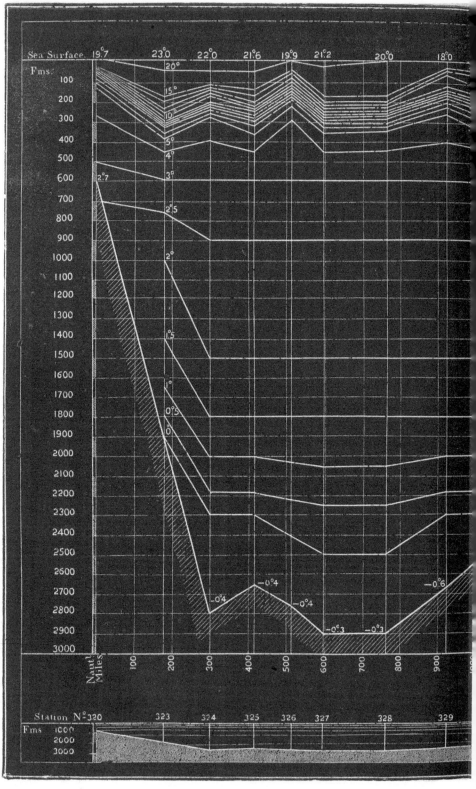

PLATE XXXVII.—DIAGRAM OF THE VERTICAL DISTRIBUTION OF T.

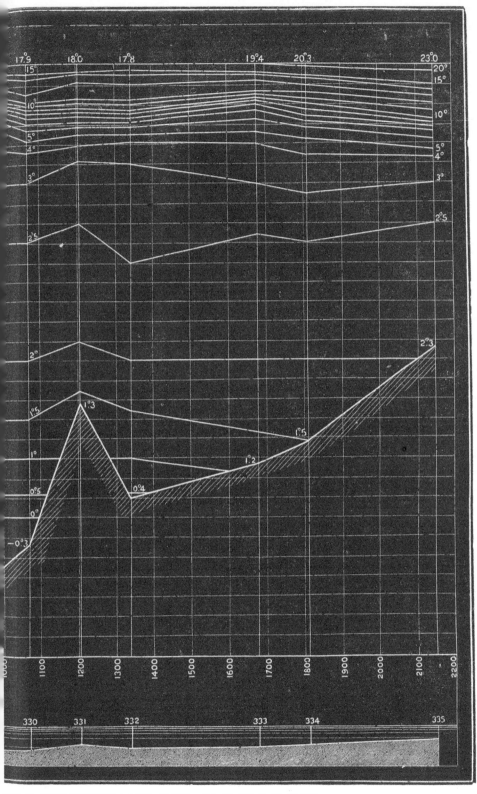

and the bottom - temperature was — 0°·4 C. The trawl-line parted near the ship in heaving in.

On the 1st of March we proceeded on our course, and on the 2nd we sounded in 2,650 fathoms with a bottom of grey mud and a bottom-temperature of — 0°·4 C. The trawl was put over, and a series of temperature-observations was taken to 1,500 fathoms. This sounding is very instructive ; the isotherm of 3° C. is found at 600 fathoms, so that we have a mass of water at a lower temperature than 3° C. 2,000 fathoms in thickness. 2°·5 C. occurs at 1,900 fathoms, and zero at 2,400. A very marked hump on the curve which extends from a depth of 125 fathoms to a depth of 255 fathoms and corresponds with the wide spaces between the isotherms of 15°C. and 6°C., evidently indicates the position and volume of the Brazil current, the southern deflection of the equatorial current after its bifurcation at Cape St. Roque. The trawl came up containing an unusually large number of organisms for this depth ; including two specimens of an undescribed species of *Euplectella,* some corals, several echinoderms illustrating three of the orders, some beautiful examples of a species of *Stylifer* commensal on one of the holo-thurians, and several fishes.

Next day we sounded in 2,775 fathoms and took temperature - soundings. This series presented a marked difference from that of the previous day. All the isotherms from that of 1°·5 C. had risen pal-pably, most of them, between 100 and 200 fathoms. Even the surface participated in the fall of tem-perature, having sunk from 21°·6C. to 19°·9C. This is evidently a space in the Brazil current occupied

by cold water, like the peculiar cold interdigitations
which are so marked in the Gulf Stream. The
position of this sounding was Lat. 37° 3′ S., Long.
44° 17′ W. A serial temperature-sounding on the
following day, at a distance of 80 miles to the
eastward, where the depth was 2,900 fathoms and
the bottom-temperature — 0°·3 C., showed by the
sinking of all the isotherms that we had again
entered the normal flow of the Brazil current.

On the 6th of March it was blowing hard from the
south-west with a heavy sea. We sounded in 2,000
fathoms, with a bottom of grey mud, and a bottom-
temperature of — 0°·3 C.; but the weather was too
boisterous to admit of a serial temperature-sounding.
On the 7th the sea was more moderate, and we
sounded in 2,675 fathoms, with a bottom-temperature
of — 0°·6 C, and took a series of temperatures. The
bottom was again a fine grey or slightly reddish
mud, almost free from calcic carbonate; samples of
water were obtained for specific-gravity determinations
and analysis down to 2,000 fathoms.

On the 8th of March we sounded in 2,440 fathoms,
with a bottom of light-red mud and a bottom-
temperature of — 0°·3 C.; and on the 9th, somewhat
to our surprise, we sounded in 1,715 fathoms with
a bottom of globigerina-ooze and a temperature of
1°·3 C. The sea was heavy, and trawling operations
consequently difficult; the trawl was lowered,
however, on account of the remarkable shallowness
of the sounding; but it unfortunately came up foul
and the observation was lost. It seems that this
sounding was on the central meridional rise which
separates the western from the eastern trough of the

Atlantic at a depth apparently nowhere much beyond
2,000 fathoms, near its western edge.   As usual, the
deeper isotherms showed a tendency to rise slightly
in the shallower water.

On the 10th the morning was misty and raining,
with the wind northerly, shifting to the southward
towards noon. We sounded in 2,200 fathoms, globigerina-ooze, with a bottom-temperature of + 0°·4 C.
The trawl was put over, but on being recovered it
was found to have· been down on its back; and it
contained only a few fragments of one or two sponges,
crustaceans, and echinoderms.

We ran on during the 11th and 12th, and on the
13th we sounded on globigerina-ooze at a depth of
2,025 fathoms with a bottom-temperature of 1°·2 C.
The trawl again came up empty and reversed, some
fragments adhering to the net showing that there was
a varied fauna, and that much interesting material
must have been got from a successful haul.

The position of the sounding on the 14th was Lat.
35° 45′ S., Long. 18° 3′ W.; the depth was 1,915 fathoms,
the bottom globigerina-ooze, and the bottom-temperature 1°·5 C.; the distance from Tristan d'Acunha
was 310 miles. The trawl came up again foul, with
only some fragments to indicate the presence of an
abundant fauna. As we had already crossed our
outward track in 1873, and as the temperatures at
depths uninfluenced by the changes of the seasons
seemed to verify in every way our former work, we
thought it unnecessary to go further to the eastward
on the direct line; and we took a north-easterly
course towards a point in the meridian of the Island
of Ascension, now distant from us about 1,685 miles.

We ran on next day, and on the 16th the position
of the ship was lat. 32° 24′ S., long. 15° 5′ W., 1,470
miles almost due south of Ascension, and 280 miles
north by west of Tristan d'Acunha. We sounded in
1,425 fathoms on globigerina-ooze with a bottom-
temperature of 2°·3 C. The trawl had failed so
frequently of late that we determined to send
down instead a large light dredge which we had
had made at Hong Kong for the shallow-water
sponge-producing seas of the Philippines. It came
up with scarcely any ooze and with only a small
number of animal species; but among them were
many very perfect specimens of the rare little sea-
urchin, *Salenia varispina.* It is singular that although
there were a large number of hempen-tangles attached
to the dredge and they seemed to have done their
work well, none of the Bryozoa so characteristic
of moderate depths with a bottom of globigerina-
ooze in the Atlantic were taken on this occasion.
In the evening we made sail due north.

For the next ten days, up to the 26th, we kept a
northerly course on the central ridge of the Atlantic
in soundings never exceeding 2,000 fathoms. The
bottom was globigerina-ooze except on two occasions
when the sounding-tube brought up no sample, and
the station was accordingly entered 'hard ground;'
the bottom-temperature averaged about 2° C., varying two or three tenths with differences of three or
four hundred fathoms in depth. The dredge was
lowered on the 19th in 1,240 fathoms, but it came
up empty; we made another attempt on the 21st,
and on this occasion we were more successful,
bringing up what we most wished, a supply of

globigerina-ooze for after examination. The only organism recovered was a dead whisp of *Hyalonema* spicules caught in the tangles.

On the morning of the 27th we were close to the Island of Ascension, and as we neared the land the weather became thick and heavy all round, and there was a very heavy rain-squall, which lasted some hours. It cleared off about noon, and the dark red cones and craters of the lower part of the island were visible to the north-eastward. We sounded in 425 fathoms, and put over the dredge, which was fairly successful, bringing up a large number of corals and sponges, and a number of echinoderms, including several examples of the ordinary form of *Echinus flemingii.*

I was sitting writing below as we approached the land, and did not go on deck until we had cast anchor in 11 fathoms in Clarence Bay, off 'Tartar Stairs,' the landing-pier of George Town. The sun was just setting, and the outlines and colouring had a most improbable effect; the near cones perfectly symmetrical and of a deep crimson; intermediate rough lava-masses, like cinders seen through a huge magnifying-glass, deep brown or pitch black; and in the middle 'Green Mountain' an irregular peak of grey trachyte, the grey of the rock melting into the curious blue-green of the Australian foliage above.

Ascension is certainly a strange little place. It is purely volcanic, and although there is now no sign whatever of volcanic activity, the cones of tufa are so fresh, and so defined and vivid in their different shades of brown and red, and the lava-beds are so rugged, apparently utterly unaffected by atmospheric action, that the impression is irresistible that it is a lately formed heap of cinders and ashes probably still resting upon slumbering fires. The island is irregularly oval in form, about seven and a-half miles long by six wide; the position of the central peak is lat. 7° 56′ 58″ S., long. 14° 20′ W.; it is directly in the path of the south-east trade, so that there is an exposed weather side with abrupt cliffs and pre- cipices and unsafe landing, and a lee side where there is the settlement and anchorage. As in almost all these volcanic islands in the path of constant winds, during the periods of eruption the scoriæ and ashes have been driven to leeward of the centre of action, and have produced a bank which now forms good holding anchorage-ground.

From the anchorage there is not a particle of vegetation to be seen, except the slight green tinge near the top of Green Mountain, about six miles distant; only a waste of lava and ashes, black, grey and red, rising peak after peak and ridge after ridge, until the harsh outlines and abrupt alternations of colour become somewhat softened down and mellowed in the distance. The little town is placed on a dreary plot of cinders at the end of a valley which winds up between two great cones of red ashes, and eventually reaches the foot of Green Mountain. There is a small fort mounting rather heavy guns, with a little pier beside it, where there is fair landing in moderate weather. There is, moreover, a large crane at the end of the pier, and a very slight shift of the trade-wind makes it necessary to rig a chair, or a bight of a rope as the case may be, on the chain,

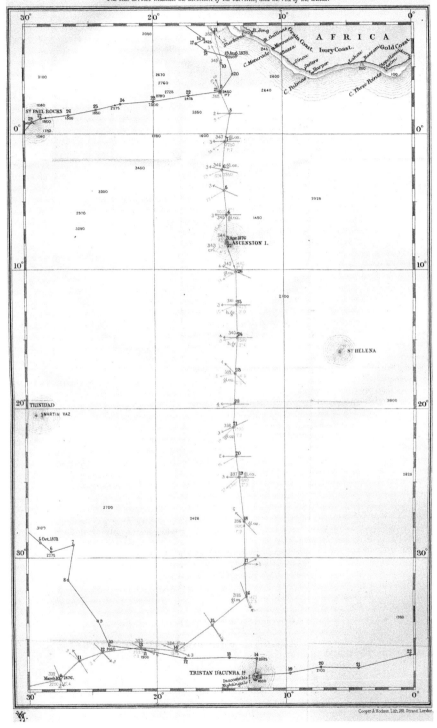

PLATE XXXVIII. *The track of the Ship from Tristan d'Acunha to Station 350.*

The blue arrows indicate the direction of the currents, and the red of the winds.

and hoist up a new arrival. A neat little church is prominent in the middle of the town, and there is a good machine-shop; a water distillery, in case of the supply on the island running short; a barrack for about a couple of hundred marines; a street of officers' quarters—neat little square houses with trim square gardens, and a full complement of ladies and healthy-looking children, and showy subtropical flowers; a commodious hospital, and a large government store.

All day one can see little parties of marines and Kroomen going to or returning from their work, or calling at certain hours at the store for rations to take home to their wives; and officers strolling about in their white tropical undress and 'puggeries,' or superintending fatigue-parties at work on the roads or in the yard.

Everything trim and neat and precise, for Ascension in one curious respect stands alone among all the isles of the sea; it, or I suppose I should say ' she,' is in commission as one of Her Majesty's ships, a tender to the ' Flora,' the guard-ship at the Cape of Good Hope; and is at present under the genial and popular command of Captain East. All the inhabitants of the island are more or less in connection with the service, and a few years ago discipline was kept up as rigidly on shore as if the island had been in truth a ship on a foreign station; smoking was allowed only at certain hours of the day, and man-of-war routine was enforced not only upon the island staff, but upon strangers also. Of late years discipline seems to have been everywhere to a certain extent relaxed; and in the Island of Ascension as elsewhere there is a great increase of community of feeling and human sympathy throughout the different grades of the service. This depends, doubtless, greatly upon the personal equation of the commandant; but not entirely so; the old oppressive system under which Ascension, in common with many other ships, suffered some years ago could scarcely exist under present conditions. Now apparently little is felt of unpleasant restriction, although the island is under military law, and everything is done in order and at the sound of the bugle. Rations are served out of food and water to every family, so much a head, the amount varying with the supply. As the island is in no sense self-supporting, nearly everything being imported, provisions are only supplied to merchant-ships in case of necessity, and at almost prohibitory rates. At noon, instead of the town-clock lagging out its twelve strokes, the workmen disperse to their mid-day spell to the sharp familiar sound of eight bells.'

The day before we arrived had been most exceptional in the experience of the station. Heavy rain had fallen, as it only knows in the tropics how to fall, for some hours continuously, too rapidly to be absorbed by the porous ashes, which soon suck up any ordinary tropical shower; and the water had rushed down the valley and swept through the settlement, committing great havoc among their neatly paved streets and squares. The torrent had rushed far out to sea red with ashes, and had carried with it quantities of cinders and lumps of pumice, some of which were still floating about on the surface.

During our stay we had a pleasant excursion up to 'Green Mountain,' where we stayed a day or two with Captain East. The road from the settlement is very good, winding up a gentle slope for the greater part of the way among the lava ridges. The whole of the lower part of the island is absolutely barren— a waste of stones with here and there a gnarled cactus-stump and a few solanaceous and portalaceous weeds, which afford scanty food to the guinea-fowl, which, at first introduced from the Cape Verde islands, have become rather numerous in the rocky valleys, and afford a good deal of very exciting, if rather break-neck sport. The most useful wild plant is the Cape gooseberry (a species of *Physalis*), which is very common, and yields an abundance of pleasant sub-acid berries. *Vinca rosea* has spread all about in its white and lilac varieties, and a tuft of its showy flowers is about the only relief to the general sterility. In a genial tropical climate, prevented from becoming insupportably hot and dry by the moisture-laden trade, and with a soil rich from the decomposition of volcanic minerals, it is wonderful what a tendency to vegetation there is. The beds are so porous that the unfrequent rain dries off at once; but even the slightest shower brings into transient blossom and beauty some little parched-up mummy of a plant in every crevice. If they could only irrigate bit by bit for a few years, till enough of vegetable soil had been accumulated to make the surface a little more compact and retentive, I am sure this wilderness would soon blossom like the rose. Natural causes will carry this out in time, and no doubt some of Captain East's remote successors in office, a few centuries hence, will be pruning their vines on the slopes of Cross Hill.

For the last mile the road zigzags up the steep slope of Green Mountain, and the whole character of the scenery suddenly changes. The clouds, driven before the south-east trade, gather and linger about the top of the mountain, and besides a frequent most refreshing mist, a reasonable amount of rain falls; not only enough to supply the requirements of the little colony on the mountain, but enough (except in exceptionally dry weather) to supply George Town also, whither it is conveyed from 'Dampier's springs' and other sources in iron pipes to a reservoir.

An area on the top of the mountain, of between four and five thousand acres, thus forms an oasis of the most delicious verdure in the middle of the desert, with a charming climate, the thermometer ranging from about 17° to 27° C.

Like Tristan d'Acunha, Ascension was first formally occupied by Great Britain as a military station in 1815, during the confinement of the Emperor Napoleon on St. Helena. After the death of Napoleon it was determined by the Admiralty to make Ascension a depôt for the refreshment of the African squadron, and a detachment of marines relieved the garrison in 1822.

The climate of Ascension is wonderfully healthy, with pure clear air, an equable temperature and a perfectly dry soil, without anything like a swamp or marsh, and with no decaying vegetation. There seem to be none of the usual endemic diseases, and patients suffering from the terrible marsh fevers of the African coast pick up rapidly the moment they are landed. For many years the chief function of Ascension was that of a sanatarium, the hospital below being filled with fever cases landed from the African ships, which were removed as soon as possible to a charming convalescent hospital on 'Green Mountain.'

On one occasion the island paid dearly for its benevolence. In the year 1823 a virulent fever was unfortunately introduced by H.M.S. 'Bann,' which carried off nearly half the population.

Of late years, for various reasons, fever has become of so much less frequent occurrence on the African station that the hospitals of Ascension are usually nearly or quite empty. The demand for fresh provisions is however an increasing one, and great care is bestowed on the cultivation of the garden and farm on Green Mountain. On a little plateau a few hundred feet below the peak there is a small barrack with a mess-room, and near it several neat detached houses with gardens, occupied by marine officers and their families; and the stables and farm-buildings. The large farm-garden—for only a few vegetables and fruits are cultivated, and these in large quantity, for the supply of the station and passing ships—is over the ridge on the south side. Sheep thrive fairly on the shoulder of the mountain, which is covered with a fine smooth sward, and planted with sheltering belts and clumps of trees chiefly introduced from Australia —*Eucalyptus* and *Acacia melanoxylon* occupying prominent places—almost like an English park. I do not know a more giddy walk anywhere than round the peak. From the considerable elevation and the small extent of the island, the descent, especially on the south side, looks almost precipitous down to the great rhythmic blue rollers breaking in cataracts of snow-white foam upon the cliffs beneath ; and as the wind is always blowing sufficiently hard to make one feel a little unsteady, it takes some little time to get sufficiently accustomed to the conditions to enjoy the view, which is certainly magnificent. The whole island, such as it is, lies at your feet like a strangely exaggerated and unskilfully coloured contour map, the great chasms and crater-valleys, even more weird and desolate, looking at them from above ; and the wide ocean of the deepest blue, flecked with white by the trade-wind, stretching round beyond to meet the sky in an unbroken and solitary circle.

The great curiosity of Ascension is ‘Wide-awake Fair ; ’ and although we had seen many such ‘fairs,’ perhaps even more wonderful, during the voyage, they are always objects of renewed interest. From Green Mountain or any of the higher peaks one can see, lying towards the shore to the right of the road from the settlement, a greyish-white patch some square miles in extent. This is a breeding-place of *Sterna fuliginosa,* called there the Wide-awake The birds are in millions, darkening the air when they are disturbed like smoke ; the eggs are excellent, somewhat like a plover’s egg in flavour. Ten thousand dozen are sometimes gathered in the breeding season in a single week, and as they are nearly as large as hen’s eggs, they are of some consideration even as an article of food.

There are at least four other species of sea-birds

abundant on the island : the frigate-bird (*Tachypetes aquila*), which causes great havoc among the young turtles as they are escaping from their nests and going down the beach to the sea; two species of *Sula*, at least two petrels, and the pretty tropic-bird (*Phäeton æthereus*), which here, as apparently all through the Atlantic, has the tail-feathers pure white. Several of these birds breed upon an outlying islet called Boatswain-bird Island.

Between Christmas and Midsummer, Ascension is constantly visited for the purpose of breeding by the common green turtle (*Chelone midas*). During that time each female is supposed to make three or four nests. The beaches in some of the bays, particularly on the west side of the island, are composed of a rough calcareous sand, made up entirely of small, smooth, rounded pieces of shell. The female turtle scrambles about 100 yards or so above high water mark, where she digs a pit eight or ten feet across by a foot or two deep, and buries in it fifty or sixty eggs, which she carefully covers over with sand. She then returns to the water till another batch of eggs is mature, when she repeats the process in another place. The young come out of the eggs in about a couple of months, and scrambling through the sand, make their way at once to the water. The females are taken by the usual operation of turning, as they are going back to the sea, and are placed in ponds into which the tide flows, below the fort at George Town. There are always a large number of the strange-looking creatures in the ponds, whence they are regularly supplied to passing men-of-war. No small turtles are ever seen. The weight of a good-sized turtle is from four to five hundred-weight. I do not think they are by any means so delicate for table use as the much smaller ones in the West Indies.

Fish are abundant round the island, and of many kinds—mullet, rock-cod, cavallas, and others. They are apparently good, for tropical fish, but of little account to those accustomed to the northern turbot and haddock. The wild quadrupeds and decapods, which may here be classed together as their habits and propensities are very similar, are rats and land crabs; both doing a great deal of damage in the gardens by destroying the roots of vegetables and fruit-trees. The rats kept out of the way during the day, but we often saw the crabs, and we were told to knock them on the head (or whatever answers that purpose) whenever we fell in with them.

I am almost disappointed that we did not see the 'rollers,' although for many reasons their occurring just at that time would have been very inconvenient. It must be a wonderful phenomenon, an enormously heavy swell arising in a perfectly calm sea, without any apparent cause, and breaking against the leeward coast of the island with almost irresistible fury. There was a slight threatening of something of the kind as we embarked with more than usual difficulty at 'Tartar Stairs' on the 2nd of April, and bade farewell to Captain East and his model colony, thinking how comparatively easy it was to make a little corner of the world tidy and comfortable and in every way respectable—if it were under discipline, and were not expected to be self-supporting.

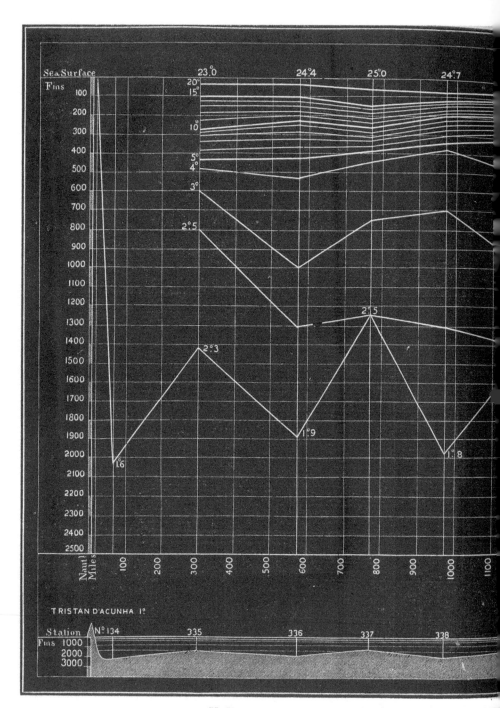

PLATE XXXIX.—DIAGRAM OF THE VERTICAL DISTRIBUTION OF

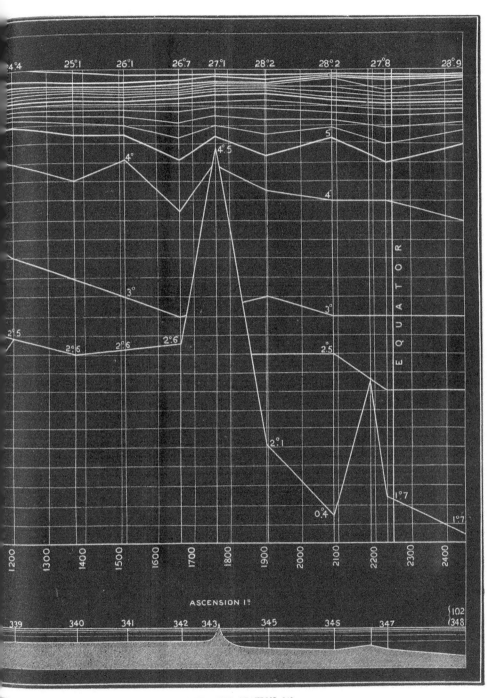

At daybreak on the 3rd we steamed out of Clarence Bay, and swung ship for errors of the compasses. In the afternoon we put over the dredge with fair result. The assemblage of animal forms was very much like that off Tristan d'Acunha, with the addition of a few more tropical species, such as *Stylaster erubescens* and a species of *Hemi-euryale*. In the evening we set sail and proceeded towards our next place of call, San Iago in the Cape Verde group.

On the 4th we sounded in 1,260 fathoms with a bottom of globigerina-ooze and a bottom temperature of 2°·1 C., and on the 6th, in 2,350 fathoms. The sounding-tube brought up a few globigerina shells and grains of manganese; the bottom temperature was 0°·4 C. The dredge was put over, but unfortunately it came up with the tangles foul and over the mouth; the number of animals was consequently small, but three fine specimens of a new species of *Porcellanaster* remarkable for a series of long spines running along the centre of the back of each ray, two samples of *Brisinga*, broken as usual, a few of *Ophioglypha bullata*, and a bryozoon had fortunately stuck to the outside of the net. From the temperature, and from the nature of the animals procured by the dredge, there could be little doubt that we had slipped off the ridge on its western side, and that the sounding was in the southern section of the western trough of the Atlantic. On the following day, after having made good 125 miles, we sounded in 2,250 fathoms with a bottom of ordinary globigerina-ooze, and a bottom-temperature of 1°·7 C. In the interval we had passed over or close to the position where the

'Gazelle' sounded in 1,640 fathoms, it therefore appears both from this and from the remarkable change in bottom-temperature, that we had crossed the ridge, and that our sounding on the 7th was in the eastern basin of the Atlantic, where all experience led us to expect a considerably higher temperature than in the south-western. We took a series of temperature-soundings down to 1,500 fathoms, and in the evening we crossed the equator for the sixth time since leaving home.

On the 9th we were close to our position on the 21st of August, 1873 (Station 102), and we put over the dredge in 2,450 fathoms. The dredge came up nearly empty, with only a small *Euplectella,* and a fragment of a large hexactinellid sponge. The bag contained a small quantity of globigerina-ooze.

For the next few days we continued our course, sometimes stopping to take temperature soundings for the first couple of hundred fathoms. The weather was fine with light northerly and north-westerly breezes, which somewhat retarded our progress. On the afternoon of the 16th we sighted the peaks of Fogo and San Iago, and after dark the lights of Porto Praya, and as the night was remarkably fine we went into Porto Praya Roads and anchored in twelve fathoms off the town. The next day we landed and revisited the 'sights' of the town and neighbourhood. In the evening we weighed and proceeded under steam and sail towards Porto Grande in San Vicente, where we anchored on the evening of the 18th.

We remained a week at Porto Grande, as the

good old ship had to be put all to rights for inspection and paying-off; and we had some pleasant rides among the hills. The town was wonderfully improved since our former visit, many new houses built, the whole place cleaned up and made more tidy, and in many places trees planted along the streets. In main features, however, San Vicente was just the same—the same barren unlovely wilderness and the same fervent heat, and the vultures still gorging themselves on the putrid flesh of the carcases half buried in the sand outside the town.

On the morning of the 26th we weighed and left Porto Grande. Towards mid-day we rounded the southern part of the island of San Antonio and shaped our course towards the Açores with a good breeze from the north. For the next week we proceeded on our course, the weather fine with light winds; and on the 3rd of May we stopped and sounded in 2,965 fathoms with a bottom of red clay, and a bottom temperature of 2°·3 C., lat. 26°16′ N., long. 33°33′ W. We were therefore, on the combined evidence of the depth and the temperature, in the prolongation to the westward of nearly the deepest portion of the eastern basin of the Atlantic. We sounded again on the 6th, lat. 32°30′ N., long. 36°8′ W., in 1,675 fathoms, with a temperature of 2°·7 C., and a bottom of pure globigerina-ooze, so that we had now passed over the edge of the trough, and were once more on the 'Dolphin Rise.' Here we fixed the position of our 354th and last deep-sea observing station.

From this point we made our way home as speedily

as we could; but our friends in England in the early part of the year 1876 may well remember the continued north-east winds which lasted until far on in the spring. These winds were dead in our teeth, and as our coal and fresh provisions began to get low, we in our weariness and impatience were driven to the verge of despair. At length, hopeless of any relenting, we resolved to go in to Vigo and get some coal and some fresh provisions, and a run on shore. As we steamed up Vigo Bay on the 20th of May the Channel Fleet under the command of Captain Beauchamp Seymour, one of the finest squadrons of ironclads ever afloat, gradually resolved itself, ship after ship, out of the mist. They were just gathering, and their tale was nearly complete, but before we left next day the fleet consisted of Her Majesty's ships 'Minotaur,' 'Iron Duke,' 'Monarch,' 'Resistance,' 'Defence,' 'Black Prince,' 'Hector,' and the despatch-boat 'Lively,' in attendance. As we rounded the stern of the 'Defence' to our anchorage, her band struck up the air 'Home, sweet Home,' and tried the nerves of some of us far more than they had ever been tried among the savages or the icebergs.

Vigo seemed very charming, but we had little time to enjoy it. We had all many friends in the fleet, and much to say and hear. While we were lying in Vigo Bay we were aware of a change of the weather, the clouds hurrying up from the south-west; so, early in the afternoon of the 21st, we weighed and proceeded to sea. Our anticipations were not disappointed; outside the bay it was blowing half a gale from the south-west, and the old 'Challenger' sped across the Bay of Biscay and up Channel at a pace very

unusual to her.　On the evening of the 23rd we passed
Ushant Light, and at 9.15 p.m. on the 24th of May,
1876, after an absence of three years and a half, we
stopped and came to an anchor in seven fathoms'
water at Spithead.

IRRIGATION ; PORTO PRAYA

APPENDIX A.

Table of Temperatures observed between the Falkland Islands and Tristan d'Acunha.

Depth in Fathoms.	Station No. 317. Lat. 48° 37′ S. Long. 55° 17′ W.	Station No. 318. Lat. 42° 32′ S. Long. 56° 27′ W.	Station No. 319. Lat. 41° 54′ S. Long. 54° 48′ W.	Station No. 320. Lat. 37° 17′ S. Long. 53° 52′ W.	Station No. 523. Lat. 35° 39′ S. Long. 50° 47′ W.	Station No. 324. Lat. 36° 9′ S. Long. 48° 22′ W.	Station No. 325. Lat. 36° 44′ S. Long. 46° 16′ W.
Surface.	8°·2C.	14°·2C.	15°·3C.	19°·7C.	23°·0C.	22°·0C.	21°·6C.
25	5·7	13·7	11·2	15·6	21·7	...	21·0
50	4·4	6·2	9·0	7·8	20·0	20·4	18·7
75	4·2	4·2	6·7	6·9	19·6	17·8	18·5
100	4·0	2·5	6·3	6·8	17·6	15·6	16·7
125	...	2·0	5·0	...	17·2	14·2	15·3
150	...	...	4·4	4·8	16·2	12·0	14·8
175	...	1·9	4·0	4·5	14·6	11·0	14·1
200	3·4	1·6	3·8	4·6	13·2	10·0	12·5
225	...	1·7	3·9	4·4	13·3	7·2	12·4
250	...	2·2	4·2	4·4	12·6	5·7	10·6
275	...	1·7	3·7	4·0	11·3	5·2	9·0
300	3·4	1·6	3·5	3·4	10·2	4·6	7·2
400	3·8	1·6	3·2	3·4	4·8	4·3	4·6
500	2·5	...	3·9	3·0	3·7	5·2	3·6
600	...	...	2·6	2·7	3·0	3·3	3·4
700	...	...	2·5	...	2·6	2·6	2·7
800	...	1·7	2·7	...	2·3	2·5	2·6
900	...	1·6	2·5	...	1·8	2·4	2·7
1000	...	1·6	2·2	...	2·4	2·7	2·5
1100	...	1·6	2·0	...	1·8	1·3	2·6
1200	...	1·6	2·2	...	1·6	2·4	2·5
1300	...	1·4	1·3	...	2·6	2·5	2·9
1400	...	1·3	2·1	...	1·6	2·1	2·4
1500	...	1·0	1·9	...	2·2	2·3	2·3
Temperature at Bottom.	1°·7	0°·3	−0°·4	2°·7	0°·0	−0°·4	−0°·4
Depth at Bottom.	1035	2040	2425	600	1900	2800	2650

Depth in Fathoms.	Station No. 326. Lat. 37° 3′ S. Long. 44° 17′ W.	Station No. 327. Lat. 36° 48′ S. Long. 42° 45′ W.	Station No. 329. Lat. 37° 31′ S. Long. 36° 7′ W.	Station No. 330. Lat. 37° 45′ S. Long. 33° 0′ W.	Station No. 331. Lat. 37° 47′ S. Long. 30° 20′ W.	Station No. 332. Lat. 37° 29′ S. Long. 27° 31′ W.	Station No. 333. Lat. 35° 36′ S. Long. 21° 12′ W.	Station No. 334. Lat. 35° 45′ S. Long. 18° 31′ W.
Surface.	19°·9 C.	21°·2 C.	18°·0 C.	17°·9 C.	18°·0 C.	17°·8 C.	19°·4 C.	20°·3 C.
25	17·6	20·0	17·7	17·2	17·0	14·9	16·8	18·1
50	14·0	19·3	13·9	15·0	14·4	13·9	14·2	15·1
75	13·2	18·7	13·1	13·9	13·3	13·6	12·9	13·6
100	11·8	17·8	12·3	13·4	13·0	13·0	12·8	13·3
125	10·2	16·6	11·8	13·1·	12·2	12·8	12·3	11·8
150	7·8	15·0	10·7	13·0	12·2	12·8	11·0	11·7
175	6·3	12·2	8·8	12·8	12·0	12·2	9·6	10·6
200	5·5	15·4	7·2	11·7	10·6	11·0	8·0	10·1
225	5·1	12·6	6·4	10·2	9·2	10·0	7·2	9·0
250	4·5	11·3	5·2	9·0	8·3	8·5	6·6	8·0
275	4·2	9·5	4·8	7·7	7·2	7·3	5·8	7·0
300	3·8	7·7	4·5	7·0	5·3	6·7	5·3	6·3
400	3·8	4·5	4·0	5·7	4·1	3·8	4·4	4·5
500	3·7	3·8	3·0	3·4	2·9	3·0	3·3	3·3
600	2·6	3·0	3·2	3·9	2·7	3·0	3·0	3·2
700	2·6	2·8	2·5	2·7	2·6	4·3	2·7	2·8
800	2·6	3·3	2·5	3·1	2·5	2·7	2·5	2·6
900	2·5	2·8	2·5	3·2	2·3	2·8	2·4	2·8
1000	2·2	2·4	2·3	2·4	2·4	2·6	2·3	2·5
1100	...	...	...	...	...	...	...	...
1200	...	...	...	...	...	...	...	...
1300	...	...	...	...	...	...	...	...
1400	...	...	...	...	...	...	...	...
1500	...	...	...	...	...	...	...	...
Temperature at Bottom.	−0°·4	−0°·3	−0°·6	−0°·3	1°·3	0°·4	1°·2	1°·5
Depth at Bottom.	2775	2900	2675	2440	1715	2200	2025	1915

APPENDIX B.

Table of Temperatures observed between Tristan d'Acunha and the Açores.

Depth in Fathoms.	Station No. 335. Lat. 32° 24′ S. Long. 13° 5′ W.	Station No. 336. Lat. 27° 54′ S. Long. 13° 13′ W.	Station No. 337. Lat. 24° 38′ S. Long. 13° 36′ W.	Station No. 338. Lat. 21° 15′ S. Long. 14° 2′ W.	Station No. 339. Lat. 17° 26′ S. Long. 13° 52′ W.	Station No. 340. Lat. 14° 33′ S. Long. 13° 42′ W.	Station No. 341. Lat. 12° 16′ S. Long. 13° 44′ W.
Surface.	23°·0 C.	24°·4 C.	25°·0 C.	24°·7 C.	24°·4 C.	25°·1 C.	26°·1 C.
25	21·0	21·0	22·2	24·1	...	...	...
50	17·8	18·3	20·7	22·1	20·6	21·8	21·9
75	16·3	16·3	18·4	19·6	. .	...	...
100	15·4	14·9	16·7	17·7	15·1	14·4	14·0
125	14·1	14·3	15·7	14·5	...	...	...
150	13·3	13·3	15·0	12·9	11·4	10·8	10·3
175	12·4	12·3	13·8	11·2	...	...	...
200	11·6	11·6	12·6	9·9	9·0	8·2	8·3
225	11·0	9·8	11·8	9·2	...	7·9	7·6
250	10·5	9·1	10·4	7·8	...	7·0	7·0
275	9·7	8·4	9·4	7·2	...	6·0	6·4
300	8·2	7·9	8·7	6·0	4·1	5·7	5·8
400	5·7	4·6	4·6	3·6	3·6	4·0	4·5
500	3·7	3·6	4·3	3·2	3·2	4·0	3·9
600	3·8	3·0	3·3	3·2	3·8	4·0	3·7
700	2·6	3·4	3·1	3·1	3·5	3·9	3·4
800	2·4	3·0	3·2	3·2	3·4	3·6	2·8
900	2·4	3·0	2·8	2·2	3·3	3·3	3·5
1000	2·6	3·8	2·8	2·9	2·9	2·8	3·2
1100	...	2·7	...	2·2	...	...	...
1200	...	3·2	...	2·7	...	...	...
1300	...	2·3	...	2·9	...	...	...
1400	...	2·3	...	2·7	...	...	...
1500	..	2·4	..	2·0	...	...	...
Temperature at Bottom.	2°·3	1°·9	2°·5	1°·8	2°·5	2°·6	3°·0
Depth at Bottom.	1425	1890	1240	1990	1415	1500	1475

Depth in Fathoms.	Station No. 342. Lat. 9° 43' S. Long. 13° 51' W.	Station No. 343. Lat. 8° 3' S. Long. 14° 27' W.	Station No. 345. Lat. 5° 45' S. Long. 14° 25' W.	Station No. 346. Lat. 2° 42' S. Long. 14° 41' W.	Station No. 347. Lat. 0° 15' S. Long. 14° 25' W.	Station No. 348. Lat. 3° 10' N. Long. 14° 51' W.	Station No. 353. Lat. 26° 21' N. Long. 33° 37' W.	Station No. 354. Lat. 32° 41' N. Long. 36° 6' W.
Surface.	26°· 7 C.	27°· 1 C.	28°· 2 C.	28°· 2 C.	27°· 8 C.	28°· 9 C.	21°· 5 C.	21°· 1 C.
25	...	...	...	22 · 2	...	...	...	19 · 1
50	21 · 9	19 · 0	22 · 1	14 · 0	18 · 4	15 · 2	20 · 0	18 · 0
75	...	...	. .	...	...	...	...	17 · 4
100	12 · 2	12 · 5	11 · 6	12 · 8	14 · 2	13 · 3	17 · 7	17 · 2
125	...	...	...	...	...	...	...	...
150	10 · 6	9 · 8	9 · 4	11 · 2	12 · 0	12 · 3	16 · 4	...
175	...	...	...	...	...	...	...	...
200	8 · 7	7 · 8	8 · 2	9 · 3	8 · 8	10 · 3	15 · 0	15 · 5
225	7 · 7	7 · 5	8 · 7	9 · 2	7 · 7	8 · 6	15 · 0	...
250	7 · 8	7 · 2	6 · 7	7 · 8	7 · 6	8 · 0	14 · 0	...
275	7 · 1	6 · 4	7 · 0	7 · 2	6 · 7	7 · 2	12 · 7	...
300	6 · 9	6 · 5	.7 · 8	6 · 2	6 · 6	6 · 5	12 · 2	11 · 9
400	6 · 0	...	5 · 6	4 · 7	6 · 0	5 · 0	9 · 5	10 · 2
500	4 · 8	...	4 · 3	3 · 8	4 · 5	4 · 3	7 · 8	7 · 4
600	3 · 4	...	3 · 6	3 · 9	5 · 9	4 · 3	6 · 2	7 · 0
700	3 · 8	...	3 · 7	4 · 1	4 · 0	4 · 2	4 · 7	5 · 8
800	3 · 2	...	3 · 8	3 · 7	4 · 2	4 · 0	5 · 2	4 · 8
900	3 · 2	...	3 · 4	4 · 3	4 · 6	3 · 7	4 · 0	4 · 3
1000·	3 · 2	...	4 · 0	3 · 6	4 · 5	3 · 4	3 · 7	4 · 0
1100	...	...	3 · 1	3 · 5	4 · 1	3 · 3	3 · 8	3 · 3
1200	...	...	3 · 0	3 · 1	3 · 4	3 · 4	2 · 1	2 · 7
1300	...	...	2 · 9	3 · 1	3 · 0	3 · 0	3 · 1	3 · 5
1400	...	...	2 · 7	2 · 7	2 · 9	2 · 5	3 · 2	3 · 6
1500	....	...	2 · 4	2 · 4	2 · 8	2 · 4	2 · 6	2 · 4
Temperature at Bottom.	2°· 6	4°· 5	2°· 1	0°· 4	1°· 7	1°·7	2°· 3	2°· 7
Depth at Bottom.	1445	425	2010	2350	2250	2250 Station 102	2965	1675

APPENDIX C.

Table of Serial Temperature-Soundings down to 200 fathoms, taken in the South and North Atlantic in the year 1876.

Depth in Fathoms.	Station No. 324. Lat. 36° 9' S. Long. 48° 22' W.	Station No. 339. Lat. 17° 26' S. Long. 13° 52' W.	Station No. 340. Lat. 14° 33' S. Long. 13° 42' W.	Station No. 341. Lat. 19° 16' S. Long. 13° 44' W.	Station No. 342. Lat. 9° 43' S. Long. 13° 51' W.	Station No. 343. Lat. 8° 3' S. Long. 14° 27' W.	Station No. 345. Lat. 5° 45' S. Long. 14° 25' W.
Surface.	22°· 0 C.	24°· 4 C.	25°· 1 C.	26°· 1 C.	26°· 7 C.	27°· 1 C.	28°· 2 C.
10	21 · 0	24 · 3	25 · 0	26 · 0	26 · 8	27 · 0	28 · 0
20	20 · 6	24 · 1	24 · 1	25 · 7	26 · 5	26 · 7	27 · 8
30	20 · 8	24 · 0	24 · 3	24 · 0	26 · 1	25 · 0	26 · 4
40	20 · 7	21 · 3	22 · 8	23 · 8	23 · 3	22 · 4	24 · 2
50	20 · 4	20 · 6	21 · 8	21 · 9	21 · 9	19 · 0	22 · 1
60	...	19 · 5	20 · 8	20 · 0	19 · 5	16 · 8	18 · 3
70	...	19 · 0	20 · 6	18 · 5	17 · 0	14 · 5	14 · 9
80	...	17 · 8	18 · 8	16 · 7	14 · 3	13 · 2	13 · 6
90	...	16 · 6	18 · 2	15 · 0	13 · 0	13 · 0	12 · 7
100	15 · 6	15 · 1	14 · 4	14 · 0	12 · 2	12 · 5	11 · 6˙
110	...	15 · 0	13 · 3	12 · 7	12 · 6	11 · 9	11 · 0
120	...	14 · 0	12 · 2	12 · 7	11 · 6	11 · 0	10 · 6
130	...	12 · 7	11 · 7	11 · 8	11 · 7	10 · 3	10 · 1
140	...	12 · 2	11 · 5	10 · 8	10 · 7	10 · 0	9 · 3
150	12 · 0	11 · 4	10 · 8	10 ·. 3	10 · 6	9 · 8	9 · 4
160	...	11 · 1	10 · 2	10 · 0	10 · 0	9 · 5	9 · 3
170	...	10 · 8	9 · 1	9 · 7	9 · 6	9 · 1	9 · 0
180	...	10 · 2	9 · 6	8 · 9	9 · 3	9 · 1	8 · 7˙
190	...	9 · 7	8 · 7	8 · 7	9 · 2	8 · 3	8 · 6
200	10 · 0	9 · 0	8 · 2	8 · 3	8 · 7	7 · 8	8 · 2

Depth in Fathoms.	Station No. 346. Lat. 2° 42' S. Long. 14° 41' W.	Station No. 347. Lat. 0° 15' S. Long. 14° 25' W.	Station No. 348. Lat. 3° 10' N. Long. 14° 51' W.	Station No. 349. Lat. 5° 28' N. Long. 14° 38' W.	Station No. 350. Lat. 7° 33' N. Long. 15° 16' W.	Station No. 351. Lat. 9° 9' N. Long. 16° 41' W.	Station No. 352 Lat. 10° 55' N. Long. 17° 46' W.	Station No. 353. Lat. 26° 21' N. Long. 33° 37' W.
Surface.	28° · 2 C.	27° · 8 C.	28° · 9 C.	28° · 6 C.	28° · 9 C.	27° · 7 C.	25° · 4 C.	21° · 5 C.
10	28 · 0	26 · 8	28 · 9	28 · 4	28 · 9	26 · 6	23 · 9	19 · 4
20	25 · 6	25 · 6	27 · 9	21 · 2	23 · 9	18 · 9	20 · 6	19 · 4
30	16 · 8	23 · 3	21 · 6	17 · 0	20 · 3	16 · 7	17 · 5	20 · 1
40	14 · 1	21 · 0	16 · 7	15 · 7	18 · 2	15 · 7	16 · 1	19 · 6
50	14 · 0	18 · 4	15 · 2	15 · 1	16 · 9	15 · 3	15 · 0	20 · 0
60	13 · 3	16 · 6	14 · 7	14 · 7	15 · 6	14 · 4	14 · 0	19 · 3
70	13 · 2	15 · 8	14 · 2	14 · 0	15 · 3	14 · 5	13 · 5	18 · 6
80	13 · 3	14 · 8	13 · 9	13 · 8	14 · 8	14 · 2	13 · 1	18 · 4
90	13 · 2	15 · 4	13 · 4	13 · 7	14 · 4	13 · 9	13 · 1	18 · 1
100	12 · 8	14 · 2	13 · 3	13 · 4	13 · 8	13 · 4	12 · 8	17 · 7
110	12 · 7	14 · 0	13 · 3	13 · 0	13 · 1	12 · 6	12 · 3	17 · 2
120	12 · 2	13 · 3	13 · 0	12 · 8	13 · 1	12 · 2	12 · 2	17 · 2
130	12 · 1	13 · 0	12 · 8	12 · 4	12 · 8	11 · 7	11 · 9	16 · 7
140	11 · 4	12 · 3	12 · 4	12 · 0	12 · 7	11 · 4	11 · 7	16 · 3
150	11 · 2	12 · 0	12 · 3	11 · 6	12 · 5	11 · 1	11 · 8	16 · 4
160	10 · 7	11 · 8	11 · 6	11 · 1	12 · 0	10 · 4	11 · 6	16 · 1
170	10 · 2	11 · 2	10 · 9	10 · 8	11 · 3	10 · 0	11 · 2	15 · 6
180	9 · 8	10 · 0	10 · 4	10 · 4	10 · 7	9 · 7	11 · 2	15 · 6
190	9 · 5	9 · 8	9 · 8	10 · 0	10 · 5	9 · 2	10 · 5	15 · 3
200	9 · 3	8 · 8	10 · 3	9 · 1	10 · 2	8 · 7	10 · 5	15 · 0

APPENDIX D.

*Specific Gravity Observations taken on the homeward voyage
between the Falkland Islands and Portsmouth.*

Date, 1876.	Latitude S.	Longitude W.	Depth of the Sea.	Depth (δ.) at which Water was taken.	Temperature (t.) at δ.	Temperature (t'.) during Observation.	Specific Gravity at (t'.) Water at 4° = 1.	Specific Gravity at 15°·56 Water at 4° = 1.	Specific Gravity at (t.) Water at 4° = 1.
			Fms.						
Feb. 7	50° 41'	56° 20'		Surface.	7°·5C.	9°·3C.	1·02637	1·02517	1·02661
8	48 37	55 17	1035	Surface.	7·7	9·9	1·02635	1·02525	1·02667
,,	...	...		25	5·7	10·4	1·02627	1·02526	1·02693
,,	...	...		50	4·4	9·9	1·02631	1·02521	1·02704
,,	...	...		100	4·0	10·0	1·02631	1·02523	1·02710
,,	...	...		200	3·4	10·1	1·02630	1·02524	1·02715
,,	...	...		300	3·4	10·1	1·02630	1·02524	1·02715
,,	...	...		400	3·4	10·1	1·02645	1·02540	1·02730
,,	...	...		Bottom.	1·7	11·5	1·02611	1·02528	1·02730
9	47 50	56 9		Surface.	10·8	10·7	1·02634	1·02535	1·02632
10	45 1	56 9		Surface.	10·5	11·7	1 02608	1·02527	1·02630
11	42 32	56 27	2040	Surface.	13·8	13·7	1·02564	1·02525	1·02562
,,	...	...		25	13·7	16·8	1·02498	1·02530	1·02569
,,	...	...		50	6·2	16·3	1·02497	1·02528	1·02689
,,	...	...		100	2·5	15·9	1·02503	1·02512	1 02708
,,	...	...		200	1·6	16·1	1·02497	1·02514	1·02714
,,	...	...		300	1·6	16·2	1·02493	1·02511	1·02711
,,	...	...		400	1·6	16·2	1·02504	1·02522	1·02722
,,	...	...		800	1·6	16·1	1·02556	1·02573	1·02771
,,	...	...		Bottom.	0·3	16·3	1·02564	1·02584	1·02782
12	41 39	54 40	2425	Surface.	14·8	16·3	1·02538	1·02559	1·02544
,,	...	...		Bottom.	-0·4	15·9	1·02544	1·02553	1·02767
13	39 33	54 20		Surface.	11·3	13·4	1·02568	1·02526	1·02612
,,	38 54	54 17		Surface.	19·2	18·6	1·02459	1·02540	1·02471
14	37 17	53 52	600	Surface.	17·6	17·8	1·02470	1·02530	1·02476
,,	...	...		25	15·6	17·1	1·02499	1·02538	1·02588
,,	...	...		50	7·8	16·9	1·02458	1·02493	1·02632
,,	...	...		100	...	16·8	1·02534	1·02567	...
,,	...	...		200	4·6	16·8	1·02503	1·02535	1·02722
,,	...	...		400	3·4	17·0	1·02500	1·02537	1·02727
,,	...	...		Bottom.	2·7	16·9	1·02510	1·02544	1·02740
15	35 4	55 6		Surface.	21·8	21·1	1·01655	Too light for	
,,	35 1	55 18		Surface.	22·2	21·6	1·01215	reduction.	
26	35 12	53 7	21	Surface.	22·0	22·3	1·02113	1·02291	1·02122
27	35 25	52 35		Surface.	23·4	23·6	1·02444	1·02667	1·02450
28	35 39	50 47	1900	Surface.	23·3	23·6	1·02460	1·02680	1·02470
,,	...	...		25	21·7	23·1	1·02475	1·02682	1·02519
,,	...	...		50	20·0	22·9	1·02494	1·02692	1·02578
,,	...	...		100	17·6	22·9	1·02471	1·02670	1·02615
,,	...	...		200	13·2	22·9	1·02428	1·02627	1·02670
,,	...	...		300	10·2	23·0	1·02385	1·02587	1·02690
,,	...	...		400	4·8	23·0	1·02450	1·02653	1·02832
,,	...	...		800	2·3	23·05	1·02461	1·02666	1·02860
,,	...	...		Bottom.	0·0	23·1	1·02445	1·02650	1·02860
29	36 9	48 22	2800	Surface.	21·9	22·0	1·02440	1·02612	1·02443
,,	...	...		Bottom.	-0·4	22·2	1·02433	1·02610	1·02820
March 1	36	47 33		Surface.	21·3	21.5	1·02492	1·02619	1·02498
,,	...	...		20	...	22·3	1·02503	1·02685	...

Date, 1876.	Latitude S.	Longitude W.	Depth of the Sea	Depth (δ.) at which Water was taken.	Temperature (t.) at δ.	Temperature (t'.) during Observation.	Specific Gravity at (t'.) Water at 4° = 1.	Specific Gravity at 15°·56 Water at 4° = 1.	Specific Gravity at (t.) Water at 4° = 1.
March 2	36° 44'	46° 16'	Fms 2650	Surface.	22°·0C.	22°·4C.	1·02499	1·02682	1·02511
,,	...	...		25	21·0	23·2	1·02475	1·02682	1·02540
,,	...	...		50	18·7	22·'4	1·02513	1·02695	1·02612
,,	...	...		100	16·7	22·6	1·02485	1·02676	1·02646
,,	...	...		200	12·5	22·55	1·02452	1·02642	1·02706
,,	...	...		300	7·2	22·7	1·02490	1·02682	1·02831
,,	...	...		400	4·6	22·9	1·02392	1·02590	1·02772
,,	...	...		800	2·6	22·8	1·02347	1·02543	1·02738
,,	...	...		Bottom.	-0·4	23·2	1·02383	1·02591	1·02805
3	37 3	44 17	2775	Surface.	20·0	21·5	1·02338	1·02494	1·02380
,,	...	...		Bottom.	-0·4	21·6	1·02420	1·02580	1·02793
4	36 48	42 45	2900	Surface.	21·2	21·5	1·02481	1·02639	1·02490
,,	...	...		Bottom.	-0·3	22·2	1·02444	1·02623	1·02836
5	37 32	42 0		Surface.	20·7	21·1	1·02462	1·02612	1·02474
6	37 38	39 36	2900	Surface.	17·6	18·8	1·02493	1·02578	1·02525
7	37 31	36 7	2675	Surface.	18·7	18·7	1·02530	1·02610	1·02610
,,	...	...		200	7·2	19·6	1·02470	1·02576	1·02725
,,	...	...		400	4·0	19·45	1 02441	1·02540	1·02727
,,	...	...		2000	1·7	19·4	1·02490	1·02588	1·02794
,,	...	...		Bottom.	-0·6	19·6	1·02477	1·02582	1·02796
8	37 45	33 0	2440	Surface.	17·9	18·5	1·02549	1·02627	1 02562
,,	...	...		Bottom.	-0·3	20·3	1·02471	1·02607	1·02820
9	37 47	30 20	1715	Surface.	18·05	18·7	1·02544	1·02628	1·02560
,,	...	...		25	17·0	20·1	1·02494	1·02612	1·02573
,,	...	...		50	14·4	20·0	1·02509	1·02624	1·02650
,,	...	...		100	13·1	20·7	1·02488	1·02625	1·02673
,,	...	...		200	10·6	17·8	1·02536	1·02595	1·02696
,,	...	...		300	5·3	19·8	1·02441	1·02551	1·02723
,,	...	...		400	4·1	17·7	1·02490	1·02547	1·02731
,,	...	..		800	2·5	18·0	1·02505	1·02569	1·02763
,,	...	...		Bottom.	1·3	20·7	1·02457	1·02591	1·02795
10	37 29	27 31	2200	Surface.	17·8	18·6	1·02531	1·02612	1·02552
,,	...	...		800	2·7	17·7	1·02502	1·02559	1·02753
,,	...	...		1400	...	17·9	1·02533	1·02594	...
,,	...	...		Bottom.	0·44	17·85	1·02525	1·02585	1·02793
11	36 34	26 1		Surface.	17·5	17·5	1·02540	1·02591	1·02540
12	35 52	24 12		Surface.	20·0	20·3	1·02481	1·02602	1·02490
13	35 36	21 12	2025	Surface.	20·1	20·7	1·02484	1·02619	1·02502
,,	...	...		25	16·8	22·9	1·02424	1·02623	1·02590
,,	...	...		50	1·2	22·4	1 02446	1·02630	1·02658
,,	...	...		100	12·8	22·4	1 02446	1·02630	1·02686
,,	...	...		200	8·0	22·5	1 02401	1 02586	1·02723
,,	...	...		300	5·3	23·5	1 02350	1·02566	1·02737
,,	...	...		400	4·4	22·6	1·02363	1·02553	1·02737
,,	...	...		800	2·6	22·4	1·02394	1·02577	1·02773
,,	...	...		Bottom.	1·2	22·6	1·02401	1·02591	1·02796
14	35 45	18 31	1915	Surface.	20·2	20·8	1·02473	1·02611	1·02491
,,	...	...		800	2·7	21·4	1·02393	1·02583	1·02778
,,	...	...		Bottom.	1·5	21·4	1·02457	1·02612	1·02814
15	34 9	15 46		Surface.	21·8	22·3	1·02459	1·02640	1·02464
16	32 24	13 5	1425	Surface.	22·9	23·3	1·02464	1·02674	1·02477
,,	...	...		25	21·0	23·3	1·02451	1·02660	1·02516
,,	...	...		50	17·8	22·8	1·02457	1·02654	1·02591
,,	...	...		100	15·4	23·0	1 02440	1·02643	1·02646
,,	...	...		200	11·8	22·9	1·02424	1·02625	1·02703
,,	...	...		300	8·6	22·8	1·02380	1·02576	1·02707
,,	...	...		400	5·7	24·0	1·02380	1·02612	1·02780
,,	...	...		800	2·5	22·8	1·02451	1·02648	1·02843
,,	...	...		Bottom.	2·3	24·1	1·02360	1·02594	1·02789
17	30 21	13 13		Surface.	24·7	24·0	1·02462	1·02695	1·02443
18	27 54	13 13	1890	Surface.	24·9	24·5	1·02457	1·02707	1·02448
,,	...	...		25	21·0	24·2	1·02442	1·02680	1·02537

Date, 1876.	Latitude S.	Longitude W.	Depth of the Sea.	Depth (δ.) at which Water was taken.	Temperature (t.) at δ.	Temperature (t'.) during Observation.	Specific Gravity at (t'.) Water at 4° = 1.	Specific Gravity at 15°·56 Water at 4° = 1.	Specific Gravity at (t). Water at 4° = 1.
			Fms.						
March 18	...	...		50	18°·3C.	23°·8C	1·02437	1·02665	1·02592
,,	...	...		100	14·9	23·9	1·02410	1·02640	1·02654
,,	...	...		200	11·4	24·1	1·02371	1·02605	1·02689
,,	...	...		300	7·6	24·0	1·02367	1·02598	1·02742
,,	...	...		400	4·7	24·2	1·02333	1·02570	1·02751
,,	...	...		800	3·0	24·3	1·02354	1·02594	1·02787
,,				Bottom.	1·9	24·5	1·02353	1·02600	1·02800
19	24°38'	13°36'	1240	Surface.	25·0	25·1	1·02450	1·02718	1·02455
,,	...	...		25	22·2	24·9	·1·02459	1·02718	1·02540
,,	...	...		50	20·7	24·75	1·02413	1·02668	1·02531
,,	..	...		100	16·8	24·6	1·02407	1·02659	1·02625
,,				Bottom.	2·5	24·3	1·02408	1·02650	1·02840
20	23 27	13 51		Surface.	24·95	25·1	1·02456	1·02722	1·02460
21	21 15	14 2	1990	Surface.	24·7	24·85	1·02504	1·02762	1·02510
,,	...	...		25	24·1	24·9	1·02472	1·02730	1·02493
,,	...	...		50	22·1	23·7	1·02488	1·02712	1·02593
,,	...	...		100	17·7	24·85	1·02418	1·02675	1·02618
,,	...	...		200	9·9	25·0	1·02343	1·02605	1·02716
,,	...	...		300	6·0	25·0	1·02396	1·02658	1·02821
,,	...	...		400	3·6	25·0	1·02400	1·02662	1·02852
22	19 55	13 56		Surface.	24·7	25·0	1·02498	1·02760	1·02507
23	17 26	13 52	1415	Surface.	24·5	24·6	1·02518	1·02768	1·02520
,,	...	...		25	24·0	24·9	1·02467	1·02726	1·02494
,,	...	...		50	20·6	25·2	1·02440	1·02710	1·02577
,,	...	...		100	15·5	25·1	1·02382	1·02649	1·02649
,,	...	...		200	9·0	24·9	1·02334	1·02592	1·02716
,,	...	...		300	4·0	24·9	1·02311	1·02570	1·02758
,,	...	...		400	3·7	25·1	1·02306	1·02570	1·02762
,,				Bottom.	2·5	25·1	1·02314	1·02580	1·02775
24	14 33	13 42	1500	Surface.	25·1	25·3	1·02492	1·02763	1·02499
,,	...	...		25	24·0	25·3	1·02470	1·02739	1·02505
,,	...	...		50	21·9	25·1	1·02471	1·02737	1·02568
,,	...	...		100	14·5	24·75	1·02387	1·02644	1·02670
,,	...	...		200	8·3	24·9	1·02483	1·02742	1·02875
,,	...	...		300	5·7	24·9	1·02327	1·02585	1·02754
,,	...	...		400	4·4	24·9	1·02386	1·02645	1·02827
,,	...	...		800	3·6	25·0	1·02311	1·02572	1·02763
,,				Bottom.	2·6	25·2	1·02341	1·02610	1·02803
25	12 29	13 44	1475	Surface.	26·2	26·1	1·02422	1·02722	1·02420
,,	12 16	13 44		25	...	26·4	1·02413	1·02720	1·02468
,,	...	...		50	21·9	26·15	1·02431	1·02731	1·02562
,,	...	...		115	...	25·15	1·02325	1·02619	1·02082
,,	...	...		200	8·3	25·9	1·02320	1·02614	1·02746
,,	...	...		300	5·8	25·9	1·02339	1·02631	1·02798
,,	...	...		400	4·5	26·0	1·02279	1·02574	1·02756
,,	...	...		800	2·8	26·05	1·02282	1·02580	1·02770
,,				Bottom.	3·0	26·45	1·02308	1·02616	1·02808
26	10 6	13 44	1445	Surface.	26·65	26·65	1·02418	1·02732	1·02418
,,	...	...		25	...	26·55	1·02408	1·02720	1·02412
,,	...	...		50	21·9	26·6	1·02385	1·02697	1·02528
,,	...	...		100	12·2	26·3	1·02309	1·02615	1·02679
,,	...	...		200	8·7	26·4	1·02319	1·02629	1·02755
,,	...	...		300	6·9	26·45	1·02309	1·02620	1·02774
,,	...	...		400	6·0	26·4	·1·02271	1·02580	1·02759
,,	...	...		900	3·2	26·45	1·02296	1·02607	1·02797
,,				Bottom.	2·6	26·7	1·02298	1·02619	1·02813
27	{ Off Ascension Island.		425	Surface.	27·1	26·8	1·02379	1·02700	1·02370
				Bottom.	4·5	25·7	1·02340	1·02622	1·02803
April 3	,,	,,		Surface.	27·7	27·4	1·02331	1·02672	1·02316
4	5 45	14 25	2010	Surface.	28·3	28·05	1·02281	1·02648	1·02272
,,	...	...		25	26·9	27·7	1·02323	1·02676	1·02350

Date, 1876.	Latitude.	Longitude W.	Depth of the Sea.	Depth (δ) at which Water was taken.	Temperature (t') at δ.	Temperature (t) during Observation.	Specific Gravity at (t') Water at 4° = 1.	Specific Gravity at 15°·56 Water at 4° = 1.	Specific Gravity at (t) Water at 4° = 1.
			Fms.						
April 4	...	...		50	22°·1C.	28°·05C.	1·02329	1·02694	1·02502
,,	...	...		100	11·6	27·6	1·02266	1·02616	1·02698
,,	...	...		200	8·3	27·6	1·02245	1·02594	1·02727
,,	...	...		300	6·6	27·6	1·02234	1·02583	1·02740
,,	...	...		400	5·5	27·7	1·02234	1·02588	1·02758
,,	...	...		1525	2·4	27·8	1·02241	1·02599	1·02793
,,	...	...		Bottom.	2·1	27·9	1·02260	1·02620	1·02814
5	4°10'S.	14°34'		Surface.	28·2	28·1	1·02248	1·02616	1·02244
6	2 42	14 41	2350	Surface.	28·2	28·2	1·02272	1·02642	1·02271
,,	...	...		25	22·2	27·8	1·02293	1·02651	1·02474
,,	...	...		50	14·0	27·9	1·02271	1·02633	1·02665
,,	...	...		100	12·8	27·8	1·02278	1·02636	1·02691
,,	...	...		200	9·5	27·8	1·02266	1·02624	1·02739
,,	...	...		300	6·5	27·7	1·02237	1·02589	1·02746
,,	...	...		400	5·4	27·8	1·02220	1·02578	1·02749
,,	...	...		800	4·0	27·9	1·02255	1·02616	1·02804
,,	...	...		1875	2·0	27·8	1·02255	1·02614	1·02814
,,	...	...		Bottom.	1·0	27·8	1·02282	1·02641	1·02848
7	0 15	14 25	2250	Surface.	27·8	27·7	1·02303	1·02657	1·02300
,,	...	...		25	25·2	27·3	1·02340	1·02683	1·02415
,,	...	...		50	18·4	27·3	1·02366	1·02706	1·02632
,,	...	...		100	14·1	27·0	1·02297	1·02625	1·02655
,,	...	...		300	6·5	26·9	1·02257	1·02580	1·02738
,,	...	...		1500	2·8	27·0	1·02283	1·02612	1·02805
,,	...	...		Bottom.	1·7	26·85	1·02281	1·02603	1·02804
8	1 30N.	14 6		Surface.	28·2	27·75	1·02271	1·02627	1·02256
9	3 10	14 51		Surface.	29·9	29·65	1·02183	1·02602	1·02179
,,	...	...		25	25·0	27·7	1·02266	1·02620	1·02358
,,	...	...		50	15·1	27·7	1·02293	1·02647	1·02655
,,	...	...		100	13·3	27·65	1·02291	1·02642	1·02699
,,	...	...		200	9·5	27·7	1·02256	1·02612	1·02727
,,	...	...		300	6·5	27·6	1·02272	1·02622	1·02780
,,	...	...		385	5·0	27·65	1·02250	1·02603	1·02780
,,	...	...		800	3·8	28·15	1·02219	1·02587	1·02777
10	5 28	14 38		Surface.	28·6	28·45	1·02256	1·02636	1·02253
,,	...	...		25	17·7	27·9	1·02287	1·02647	1·02591
,,	...	...		50	15·0	27·9	1·02281	1·02641	1·02652
,,	...	...		100	13·4	27·7	1·02284	1·02638	1·02684
,,	...	...		200	9·0	27·7	1·02266	1·02620	1·02743
,,	...	...		300	...	27·9	1·02244	1·02603	...
11	7 26	15 13		Surface.	28·5	28·3	1·02259	1·02640	1·02259
,,	7 33	15 16		25	22·0	27·65	1·02294	1·02645	1·02474
,,	...	...		50	16·9	27·65	1·02294	1·02645	1·02610
,,	...	...		100	13·8	27·7	1·02283	1·02637	1·02683
,,	...	...		300	7·1	27·55	1·02253	1·02601	1·02750
12	9 3	16 35		Surface.	27·7	27·6	1·02320	1·02671	1·02318
,,	9 9	16 41		25	17·5	26·3	1·02344	1·02652	1·02600
,,	...	...		50	15·3	26·4	1·02347	1·02653	1·02660
,,	...	...		100	13·4	26·2	1·02330	1·02633	1·02678
,,	...	...		200	8·7	26·3	1·02316	1·02624	1·02751
,,	...	...		300	6·8	26·3	1·02292	1·02598	1·02753
13	10 48	17 48		Surface.	26·0	25·9	1·02381	1·02672	1·02378
,,	10 55	17 46		25	19·0	21·45	1·02493	1·02654	1·02564
,,	...	...		50	15·0	21·55	1·02472	1·02633	1·02643
,,	...	...		100	12·8	21·5	1·02468	1·02627	1·02680
,,	...	...		200	10·5	21·55	1·02443	1·02602	1·02704
,,	...	...		300	8·0	21·55	1·02450	1·02610	1·02747
14	11 23	18 42		Surface.	23·7	23·5	1·02469	1·02687	1·02462
15	12 21	21 26		Surface.	22·8	22·7	1·02490	1·02682	1·02484
16	13 56	23 11		Surface.	23·0	22·7	1·02468	1·02660	1·02457
26	16 48	25 14		Surface.	23·0	22·8	1·02504	1·02702	1·02501
27	17 18	26 32		Surface.	22·8	22·6	1·02509	1·02700	1·02503

Date, 1876.	Lati-tude N.		Longi-tude W.		Depth of the Sea.	Depth (δ) at which Water was taken.	Temperature (t,) at δ.	Temperature (t') during Observation.	Specific Gravity at (t') Water at 4° = 1.	Specific Gravity at 15°·56 Water at 4° = 1.	Specific Gravity at (t) Water at 4° = 1.
					Fms.						
April 28	17°	47′	28°	28′		Surface.	22°· 8C	22°· 8C.	1·02499	1·02697	1·02500
29	18	20	30	10		Surface.	23 · 7	23 · 5	1·02533	1·02753	1·02530
30	20	5	30	44		Surface.	23 · 0	23 · 1	1·02555	1·02762	1·02560
May 1	21	33	31	15		Surface.	22 · 6	22 · 7	1·02578	1·02771	1·02580
2	24	0	32	38		Surface	21 · 6	21 · 7	1·02613	1·02775	1·02616
3	26	21	33	37	2965	Surface.	21 · 4	21 · 4	1·02619	1·02774	1·02618
,,	...		...			25	20 · 2	21 · 2	1·02610	1·02760	1·02639
,,	...		...			50	20 · 0	21 · 4	1·02587	1·02742	1·02628
,,	...		...			100	17 · 7	21 · 5	1·02560	1·02719	1·02661
,,	...		...			200	15 · 0	21 · 5	1·02531	1·02690	1·02702
,,	...		...			300	12 · 2	21 · 4	1·02542	1·02699	1·02668
,,	...		...			400	9 · 5	21 ·35	1·02487	1·02642	1·02759
,,	...		...			2500	2 · 5	21 · 2	1·02532	1·02682	1·02878
,,	...		...			Bottom.	2 · 3	21 · 5	1·02556	1·02714	1·02908
4	28	10	34	55		Surface.	21 · 0	21 · 2	1·02612	1·02761	1·02619
5	29	50	35	55		Surface.	20 · 7	21 · 3	1·02538	1·02741	1·02602
,,	30	20	36	6		Surface.	21 · 7	21 · 9	1·02580	1·02749	1·02587
6	32	41	36	6	1675	Surface.	21 · 2	21 · 6	1·02575	1·02735	1·02585
,,	...		...			25	19 · 1	20 · 0	1·02593	1·02708	1·02615
,,	...		...			50	18 · 0	20 · 1	1·02580	1·02700	1·02635
,,	...		...			100	17 · 2	20 · 6	1·02568	1·02701	1·02660
,,	...		...			200	15 · 5	20 · 1	1·02588	1·02706	1·02706
,,	...		...			300	11 · 9	20 ·05	1·02558	1·02675	1·02752
,,	...		...			400	10 · 2	20 · 1	1·02543	1·02663	1·02765
,,	...		...			600	7 · 0	20 · 0	1·02516	1·02632	1·02783
,,	...		...			1200	3 · 0	20 · 9	1·02558	1·02700	1·02892
,,	...		...			Bottom.	2 · 7	20 · 2	1·02544	1·02665	1·02859

APPENDIX E.

List of the Stations in the Atlantic at which Observations were
taken in the year 1876.

Station CCCXIII. January 20, 1876.—Lat. 52° 20′ S.;
Long. 68° 0′ W. Depth, 55 fathoms. Bottom temperature, 8°·8
C.; Sand.

Station CCCXIV. January 21, 1876.—Lat. 51° 36′ S.;
Long. 65° 40′ W. Depth, 70 fathoms. Bottom temperature,
7° 8 C.; Sand.

Station CCCXV. January 26, 27, 28, 1876.—Lat. 51° 40′ S;
Long. 57° 50′ W. Depth, 5 to 12 fathoms. Sand and gravel.

Station CCCXVI. February 3, 1876.—Lat 51° 32′ S.; Long.
58° 6′ W. Depth, 4 to 5 fathoms. Mud.

Station CCCXVII. February 8, 1876.—Lat. 48° 37′ S.;
Long. 55° 17′ W. Depth, 1035 fathoms. Bottom temperature,
1°.7 C. Hard ground.

Station CCCXVIII. February 11, 1876.—Lat. 42° 32′ S.;
Long. 56° 27′ W. Depth, 2040 fathoms. Bottom temperature,
0°·3 C. Grey mud.

Station CCCIX. February 12, 1876.—Lat. 41° 54′ S.; Long.
54° 48′ W. Depth, 2425 fathoms. Bottom temperature, 0°·4 C.
Grey mud.

Station CCCXX. February 14, 1876.—Lat. 37° 17′ S.; Long.
53° 52′ W. Depth, 600 fathoms. Bottom temperature, 2 ·7 C.
Hard ground.

Station CCCXXI. February 25, 1876.—Lat. 35° 2′ S.; Long.
55° 15′ W. Depth, 13 fathoms. Mud.

Station CCCXXII. February 26, 1876.—Lat. 35° 20′ S.;
Long. 53° 42′ W. Depth, 21 fathoms. Shells.

Station CCCXXIII. February 28, 1876.—Lat. 35° 39′ S. ; Long. 50° 47′ W. Depth, 1900 fathoms. Bottom temperature, 0°·0 C. Grey mud.

Station CCCXXIV. February 29, 1876.—Lat. 36° 9′ S.; Long. 48° 22′ W. Depth, 2800 fathoms. Bottom temperature, —0°·4 C. Grey mud.

Station CCCXXV. March 2, 1876.—Lat. 36° 44′ S ; Long. 46° 16′ W. Depth, 2650 fathoms. Bottom temperature, —0°·4 C. Grey mud.

Station CCCXXVI. March 3, 1876—Lat. 37° 3′ S.; Long. 44° 17′ W. Depth, 2775 fathoms. Bottom temperature, —0°·4 C. Grey mud.

Station CCCXXVII. March 4, 1876,—Lat. 36° 48′ S.; Long. 42° 45′ W. Depth, 2900 fathoms. Bottom temperature, —0°·3 C. Grey mud.

Station CCCXXVIII. March 6, 1876.—Lat. 37° 38′ S.; Long. 39° 36′ W. Depth, 2900 fathoms. Bottom temperature, —0°·3 C. Grey mud.

Station CCCXXIX. March 7, 1876.—Lat. 37° 31′ S. ; Long. 36° 7′ W. Depth, 2675 fathoms. Bottom temperature, —0°·6 C. Grey mud.

Station CCCXXX. March 8, 1876.—Lat. 37° 45′ S.; Long. 33° 0′ W. Depth, 2440 fathoms. Bottom temperature, —0°·3 C. Grey mud.

Station CCCXXXI. March 9, 1876.—Lat. 37° 47′ S. ; Long. 30° 20′ W. Depth, 1715 fathoms. Bottom temperature, 1°·3 C. Globigerina ooze.

Station CCCXXXII. March 10, 1876.—Lat. 37° 29′ S. ; Long. 27° 31′ W. Depth, 2200 fathoms. Bottom temperature, 0°·4 C. Globigerina ooze.

Station CCCXXXIII. March 13, 1876.—Lat 35° 36′ S. ; Long. 21° 12′ W. Depth, 2025 fathoms. Bottom temperature, 1°·2 C. Globigerina ooze.

Station CCCXXXIV. March 14, 1876.—Lat. 35° 45′ S. ; Long. 18° 31′ W. Depth, 1915 fathoms. Bottom temperature, 1°·5 C. Globigerina ooze.

Station CCCXXXV. March 16, 1876.—Lat. 32° 24′ S. ;

Long. 13° 5' W. Depth, 1425 fathoms. Bottom temperature, 2°·3 C. Globigerina ooze.

Station CCCXXXVI. March 18, 1876—Lat. 27° 54' S. ; Long. 13° 13' W. Depth, 1890 fathoms. Bottom temperature, 1°·9 C. Globigerina ooze.

Station CCCXXXVII. March 19, 1876.—Lat. 24° 38' S. ; Long. 13° 36' W. Depth, 1240 fathoms. Bottom temperature, 2°·5 C. Globigerina ooze.

Station CCCXXXVIII. March 21, 1876.—Lat. 21° 15' S. ; Long. 14° 2' W. Depth, 1990 fathoms. Bottom temperature, 1°·8 C. Globigerina ooze.

Station CCCXXXIX. March 23, 1876.—Lat. 17° 26' S. ; Long. 13° 52' W. Depth, 1415 fathoms. Bottom temperature, 2°·5 C. Globigerina ooze.

Station CCCXL. March 24, 1876.—Lat. 14° 33' S. ; Long. 13° 42' W. Depth, 1500 fathoms. Bottom temperature, 2°·6 C. Hard ground.

Station CCCXLI. March 25, 1876.—Lat. 12° 16' S. ; Long. 13° 44' W. Depth, 1475 fathoms. Bottom temperature, 3°·0 C. Hard ground.

Station CCCXLII. March 26, 1876.—Lat. 9° 43' S. ; Long. 13° 51' W. Depth, 1445 fathoms. Bottom temperature, 2°·6 C. Globigerina ooze.

Station CCCXLIII. March 27, 1876.—Lat. 8° 3' S. ; Long. 14° 27' W. Depth, 425 fathoms. Bottom temperature, 4°·5 C. ; coral.

Station CCCXLIV. April 3, 1876.— Off Ascension Island. Depth, 420 fathoms. Hard ground.

Station CCCXLV. April 4, 1876.—Lat. 5° 45' S. ; Long. 14° 25' W. Depth, 2010 fathoms. Bottom temperature, 2°·1 C. Globigerina ooze.

Station CCCXLVI. April 6, 1876.—Lat. 2° 42' S. ; Long. 14° 41' W. Depth, 2350 fathoms. Bottom temperature, 0°·4 C. Globigerina ooze.

Station CCCXLVII. April 7, 1876.—Lat. 0° 15' S. ; Long. 14° 25' W. Depth, 2250 fathoms. Bottom temperature, 1°·7 C. Globigerina ooze.

Station CCCXLVIII.   April 9, 1876.— Lat. 3° 10′ N. ; Long. 14° 51′ W.   Depth (see Station 102).

Station CCCXLIX.   April 10, 1876.— Lat. 5° 28′ N. ; Long. 14° 38′ W.

Station CCCL.   April 11, 1876.— Lat. 7° 33′ N ; Long. 15° 16′ W.

Station CCCLI.   April 12, 1876.— Lat. 9° 9′ N. ; Long. 16° 41′ W.

Station CCCLII.   April 13, 1376.— Lat. 10° 55′ N. ; Long. 17° 46′ W.

Station CCCLIII.   May 3, 1876.— Lat. 26° 21′ N. ; Long. 33° 37′ W.   Depth, 2965 fathoms.   Bottom temperature, 2°·3 C. Red clay.

Station CCCLIV.   May 6, 1876.— Lat. 32° 41 N. ; Long. 36° 6′ W.   Depth, 1675 fathoms.   Bottom temperature, 2°· 7 C. Globigerina ooze.

CHAPTER V.

GENERAL CONCLUSIONS.

The Contour of the Bed of the Atlantic.—The Atlantic Ocean divided by a series of Ridges into three Basins.—The Nature of the Bottom.—Pelagic Foraminifera.—*Hastigerina Murrayi.*—Volcanic *débris.*—Products of the Decomposition of Pumice.—The Distribution of Ocean Temperature. — Laws regulating the Movements of the Upper Layers of the Atlantic.— Corrections of Six's Thermometers.—Laws regulating the Movement of Water in the Depths of the Atlantic.—The doctrine of 'Continuous Barriers.'—The Distribution and Nature of the Deep Sea Fauna.—The Universal Distribution of Living Beings.—Causes Modifying and Restricting the Distribution of the Higher Forms.—Relations of the Modern to the Ancient Faunæ.—The *Challengerida.*—The Density of Sea-Water.—The amount and distribution of Carbonic Acid.—Of Oxygen.

APPENDIX A.—The General Result of the Chemical and Microscopical Examination of a Series of Twenty Samples of the Bottom from the Observing Stations on the Section between Teneriffe and Sombrero.

APPENDIX B.—Table showing the relative frequency of the occurrence of the Principal Groups of Marine Animals at Fifty-two Stations at which Dredging or Trawling was carried to Depths greater than 2,000 Fathoms.

APPENDIX C.—Table showing the amount of Carbonic Acid contained in Sea-water at various Stations in Atlantic.

IT is, of course, impossible at this stage of the work, while the great bulk of the observations are

still unreduced, while the chemical analyses are
only commenced, and there has not been time
even to unpack the natural history specimens,
to give anything like a detailed account of the
additional data which have been acquired by the
'Challenger' expedition, or of their bearings upon
the various problems of physical geography. Still,
from the presence of a competent scientific staff
on board, a good deal was done during the voyage;
and certain general results were arrived at which
are of great interest even in their present crude
form.

I propose in this chapter to summarize these
results, giving briefly a general outline of the conditions as to the contour of the bottom of the ocean,
the nature of the deposits which are being laid
down on the bottom, the general distribution of
temperature, the direction and force of surface and
submarine movements of the water, and the nature
and distribution of the deep-sea fauna, according to
my present impressions; premising that these may
be modified to a certain extent by further study of
the materials in our hands: I will however confine
myself at present as far as possible to the facts
which appear to be ascertained with some approach
to certainty.

The contour of the bed of the Atlantic.—During the
first few days of our cruise we verified many previous
observations to the effect that, after passing a comparatively narrow, shallow belt, the water deepens
rather suddenly along the coasts of Europe and
North Africa, to between 2,000 and 2,500 fathoms, a
peculiar ridge, first observed by the Swedish corvette

PLATE XL.   *Meteorological Observ*

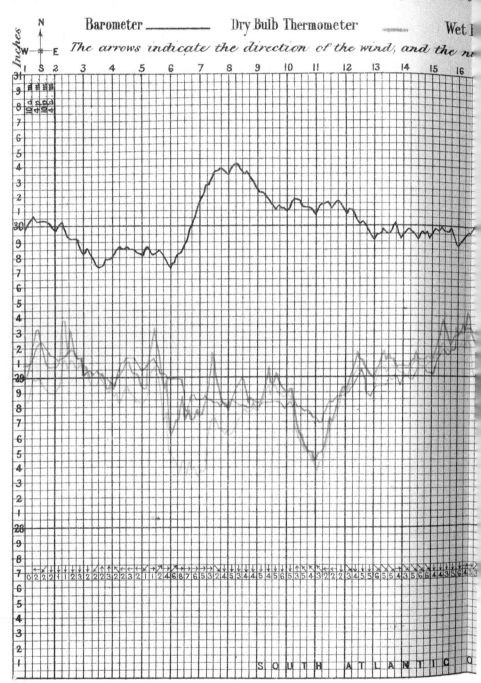

Barometer ——————   Dry Bulb Thermometer   ————   Wet l

The arrows indicate the direction of the wind, and the n

SOUTH ATLANTIC

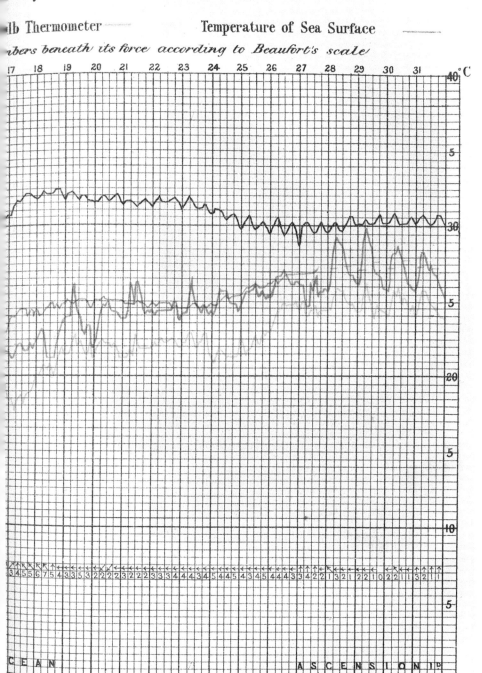

lb Thermometer   Temperature of Sea Surface

bers beneath its force according to Beaufort's scale

'Josephine,' and lately by the U.S. sloop 'Gettys-
burg,' running down in a south-westerly direction
from about Cape Roca, including the Josephine Bank
and Madeira, and giving its western boundary to a
tongue of deep water which passes in a north-
easterly direction between Madeira and the mainland
towards Cape St. Vincent.   (See Frontispiece.)

From Teneriffe westward, except at one spot about
160 miles S.W. of the island of Ferro, where we
sounded in 1,525 fathoms on the top of a ridge, the
water gradually deepened to the westward to the
depth of 3,150 fathoms at the bottom of a wide
valley which extends more than halfway across the
Atlantic.   About long. 43° W. the floor began to
rise, and at long. 44° 39' W. we sounded in 1,900
fathoms on the top of a gentle elevation.   Further to
the westward the depth again increased, and in long.
61° 28' N. we sounded in 3,050 fathoms at the bottom
of a deep western trough; the water then shoaled
rapidly up to the West Indian Islands.

On our next section from St. Thomas to Bermudas
we sounded a little to the north of the Virgin Islands
in 3,875 fathoms, the greatest depth known in the
Atlantic, and our whole course lay through a depression upwards of 2,500 fathoms deep, showing that
the western trough extended considerably to the
northward.   This western valley was again traversed
between Bermudas and the Açores, the water shallow-
ing at a distance from those islands, thus showing that
they formed the culminating points of a plateau of
considerable extent.   Between the Açores and Madeira
we recrossed the eastern valley, and our course from
Madeira to the Cape Verde Islands, and southwards

to a station in lat. 5° 48′ N., long. 14° 20′ W. lay within
it, near its eastern border. We then crossed the
valley, and in lat. 1° 22′ N., long. 26° 36′ W., we
sounded in 1,500 fathoms near the centre of the
middle ridge, and altering our course to the· south-
westward, we crossed obliquely a western depression,
with a maximum depth of about 2,500 fathoms,
between St. Paul's Rocks and Cape St. Roque. From
Bahia we crossed a western depression with a maximum depth of 3,000 fathoms, and came upon 1,900
fathoms on the central rise, a few degrees to the
westward of Tristan d'Acunha. An eastern depression with an average depth of 2,500 fathoms
extended for the greater part of the distance
between Tristan d'Acunha and the Cape of Good
Hope.

On our return voyage in 1876 we crossed the
western basin of the South Atlantic about the
parallel of 33° S. We then ran northward on the
top of the rise in the meridian of Tristan d'Acunha
and Ascension as far as the equator, and the greater
part of the remainder of our course lay nearly in
the axis of the eastern depression.

Combining our own observations with reliable
data which have been previously or subsequently
acquired, we find that the mean depth of the Atlantic
is a little over 2,000 fathoms. An elevated ridge
rising to an average height of about 1,900 fathoms
below the surface traverses the basins of the North
and South Atlantic in a meridianal direction from
Cape Farewell, probably as far south at least as
Gough Island, following roughly the outlines of the
coasts of the Old and the New Worlds.

A branch of this elevation strikes off to the south-westward about the parallel of 10° N., and connects it with the coast of South America at Cape Orange, and another branch crosses the eastern trough, joining the continent of Africa probably about the parallel of 25° S. The Atlantic Ocean is thus divided by the axial ridge and its branches into three basins: an eastern, which extends from the west of Ireland nearly to the Cape of Good Hope, with an average depth along the middle line of 2,500 fathoms; a north-western basin, occupying the great eastern bight of the American continent, with an average depth of 3,000 fathoms; and a gulf running up the coast of South America as far as Cape Orange, and open to the southward, with a mean depth of 3,000 fathoms.

The nature of the bottom.—Except in the neighbourhood of coasts, where the deposit at the bottom consists chiefly of the débris washed down by rivers, or produced by the disintegration of the rocks of the coast-line, the bed of the Atlantic, at depths between 400 and 2,000 fathoms, is covered with the now well-known calcareous deposit, the ' Globigerina-ooze,' consisting, as has been already described (Vol. I., p. 206), to a great extent of the shells, more or less broken and decomposed, of pelagic foraminifera. In the Atlantic the species producing the ooze are chiefly referable to the genera *Globigerina*, *Orbulina*, *Pulvinulina*, *Pullenia*, and *Sphæroidina*, the two latter in smaller proportions.

One very beautiful form occurs at the bottom, sparingly on account of the extreme tenuity of its shell. *Hastigerina murrayi* is very widely distri-

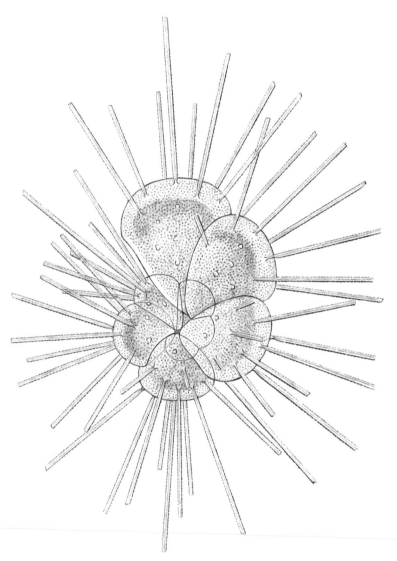

FIG. 51.—*Hastigerina murrayi*, WYVILLE THOMSON. From the surface. Fifty times the natural size.

buted on the surface of warm seas, more abundant,
however, and of larger size in the Pacific than in the
Atlantic.   The shell (Fig. 51), consists of a series of
eight or nine rapidly enlarging inflated chambers
coiled symmetrically on a plane; the shell-wall is
extremely thin, perfectly hyaline, and rather closely
perforated with large and obvious pores.   It is beset
with a comparatively small number of very large and
long spines.   The proximal portion of each spine
is formed of three laminæ, delicately serrated along
their outer edges, and their inner edges united together.   The spines, when they come near the point
of junction with the shell, are contracted to a narrow
cylindrical neck, which is attached to the shell by
a slightly expanded conical base.   The distal portion of the spine loses its three diverging laminæ,
and becomes flexible and thread-like.   The sarcode
is of a rich orange colour from included highly-
coloured oil-globules.

On one occasion in the Pacific, when Mr. Murray
was out in a boat in a dead calm collecting surface
creatures, he took gently up in a spoon a little
globular gelatinous mass with a red centre, and
transferred it to a tube.   This globule gave us
our first and last chance of seeing what a pelagic
foraminifer really is when in its full beauty.
When placed under the microscope it proved to be a
Hastigerina in a condition wholly different from anything which we had yet seen.   The spines, which
were mostly unbroken, owing to its mode of capture,
were enormously long, about fifteen times the diameter of the shell in length ; the sarcode, loaded with
its yellow oil-cells, was almost all outside the shell,

and beyond the fringe of yellow sarcode the space
between the spines to a distance of about twice the

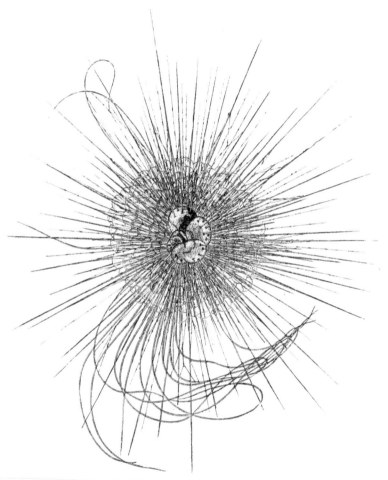

FIG. 52.—*Hastigerina murrayi*, WYVILLE THOMSON. From the surface. Ten times the natural
size.

diameter of the shell all round was completely filled
up with delicate *bullæ*, like those which we see in

some of the Radiolarians, as if the most perfectly transparent portion of the sarcode had been blown out into a delicate froth of bubbles of uniform size. Along the spines fine double threads of transparent sarcode, loaded with minute granules, coursed up one side and down the other, while between the spines independent thread-like pseudopodia ran out, some of them perfectly free, and others anastomosing with one another or joining the sarcodic sheaths of the spines, but all showing the characteristic flowing movement of living protoplasm. The woodcut (Fig. 52), excellent though it is, gives only a most imperfect idea of the complexity and the beauty of the organism with all its swimming or floating machinery in this expanded condition. We have seen nothing exactly like it in any other species. We have frequently seen *Globigerina* with spines, and the sarcode extended along them, and displaying its characteristic movements; and on one or two occasions we saw *Pulvinulinæ* with a half-contracted float, resembling partially expanded bullæ; but in all these cases the animals had been taken in the tow-net, and were greatly injured.

Everywhere in the 'globigerina-ooze,' Mr. Murray has detected, in addition to the foraminifera which make up the great part of its bulk, fragments of pumice, minute particles of felspar, particles and crystals of other minerals due to the disintegration of volcanic rocks, such as sanidine, augite, hornblende, quartz, leucite, and magnetite, and rounded concretions of a mixture of the peroxides of manganese and iron.

I have already (Vol. I. p. 223 *et seq.*) discussed very

fully the way in which, at depths over 2,000 fathoms, the carbonate of lime of the globigerina-ooze is gradually removed, the ooze becoming darker in colour and effervescing less freely with acids, until at length it gives place to a more or less homogeneous 'red-clay'; and I have referred to the relative proportions in which these two great formations occur in the Atlantic. Their distribution may be broadly defined thus: the globigerina ooze covers the ridges and the elevated plateaus, and occupies a belt at depths down to 2,000 fathoms round the shores outside the belt of shore deposits; and the 'red-clay' covers the floor of the deep depressions, the eastern, the north-western, and the south-western basins. An intermediate band of what we have called 'grey ooze' occurs in the Atlantic at depths averaging perhaps from 2,100 to 2,300 fathoms.

Over the 'red clay' area, as might have been expected from the mode of formation of the 'red clay,' the pieces of pumice and the recognizable mineral fragments were found in greater abundance; for there deposition takes place much more slowly, and foreign bodies are less readily overwhelmed and masked; so abundant are such fragments in some places that the fine amorphous matter, which may be regarded as the ultimate and universal basis of the deposit, appears to be present only in small proportion.

Mr. Murray has studied very carefully the distribution of volcanic debris over the floor of the ocean. He finds that recognizable pieces of pumice, varying from the size of a pea to that of a football,

have been dredged at eighty stations, distributed all
along our route ; and he finds them in greater abundance in the neighbourhood of volcanic centres, such
as the Açores and the Philippines, than elsewhere.
In deposits far from land they were most numerous
in the pure deep-sea clays; minute particles of felspar,
having the appearance of disintegrated pumice, were
detected in all such ocean deposits. Many of the
large pieces were much decomposed, while some were
only slightly altered ; some were coated with manganese and iron, and many appeared as a mere
nucleus, round which the manganese and iron had
aggregated. They varied greatly in structure, being
highly vesicular, or fibrous and compact, and in
colour, from white through grey or green to black.
There seemed to be every gradation from the felspathic to the extreme pyroxenic varieties.

Mr. Murray believes that all the pieces of pumice
which we find at the bottom of the sea have been
formed by sub-aërial volcanic action. Some of them
may have fallen upon the sea; but the great majority
seem to have fallen on land, and been subsequently
washed and floated out to sea by rains and rivers.
After floating about for a longer or shorter time
they have become water-logged and have sunk to
the bottom. Both in the North Atlantic and in
the Pacific small pieces of pumice were several
times taken on the surface of the ocean by means
of the tow-net. Over the surface of some of these,
serpulæ and algæ were growing, and crystals of
sanidine projected, or were imbedded in the felspar. During our visit to Ascension there was a
very heavy fall of rain, such as had not been

experienced by the inhabitants for many years. For several days after, many pieces of scoriæ, cinders, and the like were noticed floating about on the surface of the sea near the island. Such fragments may be transported to great distances by currents.

On the shores of Bermudas, where the rock is composed of blown calcareous sand, we picked up fragments of travelled volcanic rocks. The same observation was made by General Nelson at the Bahamas. Mr. Darwin observed pieces of pumice on the shore of Patagonia, and Professor L. Agassiz and his companions noticed them on the reefs of Brazil. During a recent eruption in Iceland, the ferry of a river is said to have been blocked for several days by the large quantity of pumice floating down the river and out to sea.

Near volcanic centres, and sometimes at great distances from land, we find much volcanic matter in a very fine state of division at the bottom of the sea. This consists mainly of minute particles of felspar, hornblende, augite, olivine, magnetite, and other volcanic minerals. These particles may probably have been in many cases carried to the areas where they are found by winds in the form known as volcanic dust or ashes. Mr. Murray examined a packet sent to me by Sir Rawson Rawson, of volcanic ashes which fell at Barbadoes in 1812, after an eruption on the Island of St. Vincent, a hundred and sixty miles distant; and he found them to be made up of particles similar to those to which I have referred.

The clay which covers, broadly speaking, the

bottom of the sea at depths greater than 2,000 fathoms, Mr. Murray considers to be produced, as we know most other clays to be, by the decomposition of felspathic minerals; and I now believe that he is in the main right. I cannot, however, doubt that were pumice and other volcanic products entirely absent, there would still be an impalpable rain over the ocean floor of the mineral matter which we know must be set free, and must enter into more stable combinations, through the decomposition of the multitudes of organized beings which swarm in the successive layers of the sea, and I am still inclined to refer to this source a great part of the molecular matter which always forms a considerable part of a red-clay microscopic preparation.

There is great difficulty in pointing out rocks belonging to any of the past geological periods which correspond entirely, whether in chemical composition or in structure, with the beds now in process of formation at the bottom of the ocean. There seems every reason to believe that the rocks of the Mezozoic and Cainozoic series, at all events, were formed in comparatively shallow water, and after the prominent features at present existing had been stamped upon the contour of the earth's crust; and consequently that none of these have the essential characters of deep-sea deposits. I imagine, however, that the limestone which would be the result of the elevation and slight metamorphosis of a mass of ' globigerina ooze ' would resemble very closely a bed of grey chalk; and that an enormous accumulation of ' red clay ' might in time under similar circumstances come to be very like one of the palæozoic schists, such, for

example, as the Cambrian schist with Oldhamia
and worm-tracks at Bray Head. It is a very difficult question, however, and one on which I shall
offer no opinion until we have very much more
complete data from comparative microscopical
examination and chemical analysis.

The Distribution of Ocean Temperature.—Throughout the whole of the Atlantic the water is warmest
at the surface; from the surface it cools rapidly
for the first hundred fathoms or so, it then
cools more slowly down to five or six hundred
fathoms, and then extremely slowly either to the
bottom or to a certain point, from which it maintains a uniform or nearly uniform temperature to
the bottom.

A glance at a series of temperature sections such
as those represented in Plates V., IX., XVI., XX.,
XXII., and XXVIII. gives the impression that a
generally uniform temperature is maintained by
a belt of water at a depth of from seven to eight
hundred fathoms, and that this belt separates two
bodies of water which are under essentially different
conditions; above, the vertical distribution of temperature differs greatly in different localities, while
below the uniform belt there is a slow and gradual
cooling, which also differs both in rate and in amount
in different localities, but in another way. These
variations in temperature, whether in the superficial
layers or in the deeper, are undoubtedly in all cases
connected with currents or movements of the water,
and may be regarded as evidences of portions, modified
by various causes, of a general system of circulation
of the water of the ocean.

The movements of surface water may usually be determined with considerable precision by a comparison at the end of a given time of the apparent course of a ship and her position by dead reckoning, with her actual position by observation. The rate and direction of a surface current may also be ascertained by getting in some way a fixed point, by anchoring a boat for instance, and observing and timing the course of a body floating past it. Neither of these methods can be satisfactorily applied to deep-sea currents; indeed it seems probable that the movements of masses of underlying water are so slow, that, even if we had some feasible method of observation, the indications of movement within a limited period would be too slight to be measured with any degree of accuracy.

We cannot therefore measure these currents directly, but we have in the thermometer an indirect means of ascertaining their existence, their volume, and approximately their direction. Water is a very bad conductor of heat, and consequently a body of water at a given temperature passing into a region where the temperature conditions are different, retains for a long time, without much change, the temperature of the place where its temperature was acquired. To take an example: the bottom temperature near Fernando Noronha, almost under the equator, is $0°·2$ C., close upon the freezing-point; it is obvious that this temperature was not acquired at the equator, where the mean annual temperature of the surface-layer of the water is $21°$ C., and we may take the mean normal temperature of the crust of the earth as not lower

at all events than 8° C. The water must therefore
have come from a place where the conditions were
such as to impart to it a freezing temperature ; and
not only must it have come from such a place, but
it must be continually renewed, however slowly, for
otherwise its temperature would gradually rise by
conduction and mixture. Across the whole of the
North Atlantic the bottom-temperature is considerably higher, so that the cold water cannot be coming
from that direction ; on the other hand, we can trace
a band of water at a like temperature at nearly the
same depth continuously to the Antarctic Sea, where
the conditions are normally such as to impart to it
its low temperature. There seems therefore to be no
room for doubt that the cold water is welling up
into the Atlantic from the Southern Sea ; we shall,
however, discuss this more fully hereafter.

The investigation, by this indirect method, of the
movements of the water of the ocean, was one of
the points to which our attention was very specially
directed ; and it was prosecuted throughout the voyage
with great care. The method of taking temperature
sections was first systematically employed, so far
as I am aware, by the American Coast Survey in
their examination of the Gulf-stream, and some
modifications, extending its use to deep water, were
devised during the cruises of the ' Lightning ' and
' Porcupine '; and the instructions to the ' Challenger '
were chiefly based on our experience in the preliminary trips. (See ' The Depths of the Sea,' p. 284
et seq.)

The observing stations were fixed as nearly as
possible in a straight line, if possible either meridianal

or on a parallel of latitude; the bottom-temperature was carefully determined by the mean of two observations; a string of thermometers was then sent down, in detachments to avoid the risk of too great a loss in case of an accident, at intervals of 100 fathoms to within 100 fathoms of the bottom, or more usually to a depth of 1,500 fathoms, considerably beyond the uniform layer.

Such observations gave us a very fair idea of the distribution of temperature along a section, and the general course of groups of lines joining points of equal temperature along the section gave very delicate indications of any general rise or fall. The word 'isotherm' having been hitherto so specially appropriated to lines passing through places of equal temperature on the surface of the earth, I have found it convenient, in considering these questions of ocean-temperature, to use the terms 'isothermobath' and 'isobathytherm;' the former to indicate a line drawn through points of equal temperature in a vertical section, and the latter a line drawn through points of equal depth, at which a given temperature occurs. Isothermobaths are shown in schemes of a vertical section such as those in Plates V., IX., XI., &c.; isobathytherms are of course projected on the surface of the globe. All the temperature observations have been made with the modification of Six's registering instrument known under the name of the Miller-Casella thermometer, and this instrument, although a great advance upon any other hitherto constructed, is essentially uncertain and liable to error from various causes; thus even a slight jerk causes the index to move slightly either up or down, and an observation is in this way very frequently vitiated. In almost every serial temperature-sounding one or two of the thermometers were evidently adrift from some such cause. There was an excellent proof that these eccentricities did not always depend upon differences of temperature. Very frequently, especially at considerable depths, where the differences were very slight, thermometers sent to greater depths gave indications higher than those above them; there may be no absolute reason why underlying water might not in some cases have a temperature higher than that of the layers above it, but the thermometer is not constructed to show such an anomaly; having once registered its minimum, it has no power of amendment.

I have no hesitation therefore in saying that any single indication with a thermometer on Six's principle is not trustworthy, and that a fact in temperature distribution can only be established by a series of corroborative determinations.

Although the gross errors to which an unprotected thermometer is liable from pressure may be said to be got rid of by the addition of the outer shell, a certain amount of error in the same direction still remains, probably from a slight compression of the unprotected parts of the tube. This error, which is one of slight excess, although for practical purposes it might perhaps be safely regarded as the same for all thermometers, is in detail special to each instrument, and all our thermometers were tested by Captain Tizard, and their individual errors tabulated for every 100 fathoms.

The following table, which is given as an example, is in Fahrenheit degrees.

Number of Thermometer.	Correction for 100 fathoms.	For 500 fathoms.	For 1,000 fathoms.	For 1,500 fathoms.	For 2,000 fathoms.	For 2,500 fathoms.	For 3,000 fathoms.
O 1	0	0·2	0·5	0·7	0·9	1·1	1·4
O 2	0	0·2	0·4	0·6	0·8	1·0	1·2
O 3	0	0·2	0·4	0·6	0·8	1·0	1·2
O 4	0	0·4	0·7	0·8	0·9	1·1	1·1
O 5	0	0·3	0·6	0·8	0·9	1·1	1·2
O 6	0	0·3	0·6	0·8	0·9	1·1	1·2
O 7	0	0·2	0·4	0·6	0·7	0·9	1·1
O 8	0	0·2	0·4	0·6	0·8	1·0	1·2

These particular thermometers were part of a batch sent out to us late in the cruise, specially strengthened, and certainly of a better construction than those which we had had before. By testing a large series of the earlier instruments in a Bramah's press, Captain Davis had come to the conclusion that when subjected to a pressure corresponding to a depth of 2,000 fathoms, they gave, broadly, a uniform error of 1° ·4 F. in excess, and that this correction might be applied proportionally to the depth at which the observation is taken, *i.e.* 0° ·7 F. for every 100 fathoms. This may probably hold as a rough rule for ordinary instruments, where absolute accuracy is not required.

On reconsidering this matter since our return home, a doubt has arisen whether we were justified in applying to the minimum side of the thermometer these corrections on the scale prepared by Captain Davis, and a new set of experiments has been commenced at pressures up to three to four tons on the square inch.

This last class of errors may seem very trivial, but there are cases, where questions of special delicacy arise, in which they may assume considerable importance. Throughout the ocean generally, at all events between the two polar circles, the temperature of the ocean may be said as a rule to sink from the surface to the bottom. There are many places however where this gradual sinking appears to be arrested at a certain point, from which the temperature remains uniform to the bottom. Frequently the temperature as recorded by the thermometer reaches a minimum at a depth of 1,800 or 2,000 fathoms: this is the case, for example, throughout the greater part of the Atlantic, and there is little doubt that the result is in the main correct, and can be accounted for by the action of a very simple law; but if the temperature remained *exactly* the same the application of this ultimate correction to depths from 2,000 down to 3,000 fathoms would cause the thermometer to appear to rise sensibly. This certainly is not generally the case, or it would have come out in the large number of observations which have been made under circumstances where such a result might have been expected; and therefore I think we must conclude that in all the great ocean basins, from some cause or other, there is a very slight fall of temperature to the very bottom.

In order to eliminate as far as possible from the results of our serial temperature-soundings errors depending upon irregularities in the action of the thermometers, it has been found necessary in all

cases, instead of trusting to their individual indications, to construct a free-hand curve for each
series, and to take the indications from the curve.

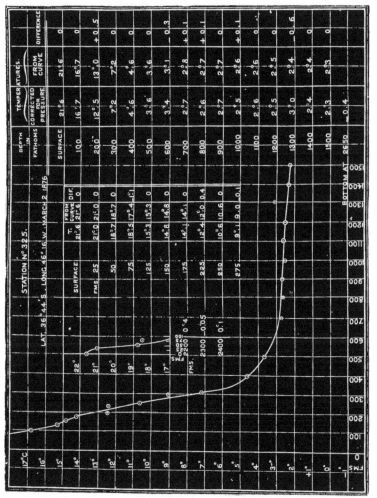

FIG. 53.—A page of the Temperature 'Curve-book,' for Station 325

If the readings of the thermometer are plotted to
scale, and if we attempt to construct such a symmetrical curve as that represented in Fig. 53, a page of

x 2

the Curve-book selected at random as an example, the curve naturally passes through the greater number of the plotted points, leaving out one or two at a greater or less distance at either side.   Where many of the thermometers are astray, as not unfrequently happens when the serial sounding is taken in heavy weather, this process requires to be performed with some judgment, and is liable to a certain amount of error ; but it is wonderful in a series of such curves how strong the internal evidence is of their accuracy.   A certain marked temperature phenomenon, for example, is indicated in a certain locality by an irregularity in the curve, and as we recede from the cause of disturbance, the irregularity gradually dies out, to be replaced very probably by an irregularity due to some other cause.   This is well shown in the curves representing the gradual change of temperature from west to east in the North Atlantic and the North Pacific (Vol. I., Figs. 100 and 101, pp. 392 and 394). The temperatures used in the text and in the temperature sections are taken from such curves ; in the Tables in the Appendices the temperatures are given as they were read from the thermometers, after applying the known corrections for pressure and error of zero point, in order that the actual data from which the curves were constructed might be supplied.   This will explain the discrepancies which frequently occur between the temperatures referred to in the text and those given in the Tables.

Referring in the first place to the distribution of temperature in the layer extending from the belt of comparatively constant temperature to the surface, on our first section from Teneriffe to Sombrero, we found the temperatures below the 100 fathom line very uniform, the upper isothermobaths crowding together a little, and the lower slightly rising to the westward. The main feature in the section was the steady increase to the westward of the temperature of the surface layer, the isothermobaths of 19°, 20°, 21°, 22°, 23°, and 24° C. being added in regular succession. This was due chiefly to the southward direction of the section, partly to the advance of the season, and partly to the westward determination of the warm surface water.

From St. Thomas to Bermudas the surface temperature naturally fell gradually, the lower lines remaining pretty steady; but at station 28, lat. 24° 39′ N., long. 65° 25′ W., a very marked widening of the space between the isothermobaths of 18° and 19° C. was observed, and, farther on, the whole of the space between the lines of 16° and 19° C. became abnormally expanded, indicating the presence of a layer of water 200 fathoms thick, considerably above its normal temperature, lying between the 100 and the 300 fathom lines. This warm band appeared again to the north of Bermudas, and on our northwesterly course towards Sandy Hook (Plate XI.) it maintained its volume and position to lat. 36° 23′ N., long. 71° 51′ W., when it came to the surface or became merged in the phenomena of the Gulf-stream.

We next crossed the Gulf-stream, of which I have nothing further to say than that all we saw confirmed our previous convictions as to its cause and its effects; and we sounded in the Labrador current, the local and most insignificant return stream from the Arctic Sea.

On our return from Halifax to Bermudas (Plate IX.) we again encountered the warm band at station 52, lat. 39° 44′ N., long. 63° 22′ W., and traced it all the way to the island.   To the east of Bermudas it again made its appearance on our section from Bermudas to the Açores (Plate XVI.), and maintained its volume to station 70, lat. 38°-25′ N., long. 35° 50′ W., where it became less definite, and then thinned out, while at the same time the lower isothermobaths began to dip down and to separate, indicating an enormous accumulation of super-heated water, occupying depths between 400 and 1,000 fathoms.   This condition continued up to the Island of Madeira; we had already established that it extends as far north as the Bay of Biscay.

South of Madeira the deep warm band steadily narrowed up to the Cape Verde Islands; and after we passed the Bijouga Islands and were in the full tide of the Guinea current, the isothermobaths had gathered up to the surface, the line of 5° C. being at 300 fathoms, and reducing the warm water to a mere superficial layer.   The next section, from station 102 to Pernambuco (Plate XXII.) was nearly equatorial, and the same singular condition was maintained throughout—an exceedingly rapid fall for the first 300 fathoms to a temperature of about 5° C., with an underlying mass of cold water of vast thickness.

Shortly after leaving Bahia we crossed the warm surface water of the Brazil current, and as the first part of our course, as far as Tristan d'Acunha, then lay in a south-easterly direction, the surface temperature of course steadily declined, the isothermobaths between 10° and 4° C. maintaining their previous course, crowded together between the depths of 100 and 400 fathoms (Plate XXVIII.). From Tristan d'Acunha the temperature for the first 600 fathoms remained very uniform in its rate of cooling until we were within little more than twenty miles of the Cape of Good Hope, when a sudden rise in all the higher temperatures told us that we had entered the westward loop of the Agulhas current.

In the southern summer of 1876, on our course from Monte Video to Tristan d'Acunha, for the first 900 miles we traversed the southern extension of the Brazil current, which depressed the isothermobaths of 15° C. to a depth of nearly 200 fathoms, with some cool interdigitations (Pl. XXXVII.), and the temperature remained very equable for the remainder of the section, the spaces between the higher isothermobaths widening a little to the eastward. On the meridianal section from Tristan d'Acunha to the equator, the isothermobaths between 5° and 20° C. altered very slightly in position; the surface layers of course became steadily and rapidly warmer.

Bearing in mind that at a certain depth below the surface, varying only slightly in different regions, there is a thick belt of water at a pretty nearly uniform temperature from 4° to 5° C., it is evident that the much higher temperature of the surface-layers must be due, for each position, directly or indirectly to the heat of the sun. Normally the surface-temperature would attain its maximum near the equator, and would decrease uniformly towards the poles; and the very abnormal distribution of temperature which actually exists must depend upon some disturbing cause or causes. That several such causes come into play, and many complicated combinations of these causes, there appears to be little doubt; but one disturbing cause seems to be so paramount, so sufficient in itself to account for the observed phenomena, that I do not think it necessary in this preliminary sketch to pursue the inquiry beyond it.

The permanent winds, blowing eternally in one direction where the water is hottest, send the heated surface-water in a constant stream to the westward. This 'equatorial current,' impinging upon the coast of South America about Cape St. Bogue, splits in two; a considerable portion of the northern branch coursing round the Gulf of Mexico, and becoming contracted and condensed by the Strait of Florida, makes itself manifest as the celebrated Gulf-stream, while the remainder, moving outside the islands in a gentler and less obvious current, spreads over the great bight between North and South America, and gives an indication of its presence in the high thermometer-readings round Bermudas, and westward to the Açores. The cause of the second and deeper hump on the temperature curves (Vol. I. Fig. 100), in a section between Bermudas and the coast of Europe, is perhaps not so evident; the explanation which I have suggested elsewhere is that the warm water of the Gulf-stream, forced to the eastward by its high initial velocity, and thus accumulated at the head of the Atlantic whence it has no free egress, becomes 'banked down,' and the warm stratum abnormally thickened against the coast of Western Europe. Some ingenious theories, depending upon changes of density produced by evaporation, and changes of density combined with changes of temperature, have been proposed to account for the great accumulation of water of abnormally high specific gravity, and at an abnormally high temperature in the North Atlantic; but these do not seem to be satisfactory, and as they can only be supposed to act, at most, as very subordinate auxiliaries to the wind circulation, they sink in importance into the category of questions of detail.

The branch of the equatorial current deflected to the southward of Cape St. Roque, passes down as the Brazil current, parallel with the coast of South America. In its southward extension it finds no barrier corresponding to that which circumscribes and moulds the northern branch; gradually widening out and becoming less defined, at the same time acquiring a sufficient easterly deflection to keep it out from the coast, it is at length almost merged in the great easterly drift-current which sweeps round the world, occupying a belt 600 to 1,000 miles broad in the Southern Sea. But while the greater part of the Brazil current is thus merged, it is not entirely lost; for at its point of junction with the drift-current of the westerlies all the upper isobathytherms are slightly deflected to the south, and opposite the point where this deflection occurs there is comparatively open sea far to the southward, and a penetrable notch in the southern pack; taking advantage of this, Weddell in 1829, and Ross in 1843, reached the parallels of 14° 14′ and 71° 30′ S. respectively, between the meridians of 15° and 30° W. The same thing occurs with regard to the Agulhas current and the East Australian current; but the case of the Brazilian current is a little more complicated than that of the other two, for there is high and extensive land between the Meridians of 55° and 65° west, in 65° south latitude, and the warm current, already led far to the southward by the American coast, appears to bifurcate upon Graham Land, and to produce another bight in 90° west longitude, a little to the west of the southern point of South America. In this bight, Cook in 1771, and Bellingshausen in 1821, pushed nearly to the seventieth parallel of south latitude.

I have already referred (Vol. II. p. 78), to the principal temperature phenomenon of the eastern portion of the South Atlantic—the equatorial counter-current, and its extension as the Guinea current. The cause of the counter-current to the eastward in the zone of calms is somewhat obscure, as the only obvious explanation, that it is a current in an opposite direction induced in the space between the current of the north-east and south-east trades to supply the water removed by them, seems scarcely sufficient to account for its volume and permanence.

The comparative thinness of the belt of warm surface-water in the equatorial region is at first sight remarkable, and has given rise to a good deal of speculation; but it will be seen by comparing the distribution of temperature at Station 112 (Fig. 54), nearly on the line, with that at Station 327 (Fig. 56), in the latitude of Tristan d'Acunha, that the positions of the isothermobaths of 4° and 5° C. are nearly the same; the slight difference apparently depends upon the latter station being within the influence of the Brazil current. The phenomenon is thus essentially a continuation to the north of the equator of southern conditions, and the small effect of the vertical sun in raising the temperature to any depth below the surface is doubtless due to the removal of the heated layer as soon as it is formed by the trade-winds and their counter-currents, and to the rapid abstraction of heat in the formation of watery vapour.

One of the best-marked and most important phenomena of the distribution of temperature in the upper layers of the Atlantic is the steady increase in the volume of warm water from the south northwards. For example, between Montevideo and Tristan d'Acunha we find the isothermobath of 7° C. at an average depth of about 250 fathoms, along the equator at under 300 fathoms, between Teneriffe and Sombrero it occurs at a depth of 500 fathoms, and between Bermudas and Madeira at about 600 fathoms ; the principal accumulation of warm water, at depths below 400 fathoms in the North Atlantic, is to the eastward.

We now pass to the more difficult problem of the distribution of temperature in the mass of water filling up the trough of the Atlantic beneath the uniform belt. The isothermobath of 3° C. may, perhaps, be regarded as the first line decidedly within the upper boundary of the cold water, and we learn something by observing its position. In the most northern cross-section, between Bermudas and Madeira, it occurs at a depth of from 1,000 to 1,200 fathoms below the surface. In the next cross-section, from Teneriffe to Sombrero, it has nearly the same position, becoming a little deeper towards the eastward. In the next section, along the equator, it is at a depth of from 1,000 to 1,100 fathoms, nearly as before. Between San Salvador and the Cape of Good Hope it rises to a mean depth of 600 fathoms, and between the Falkland Islands and Tristan d'Acunha it is at a depth of from 500 to 600 fathoms. The broad fact thus becomes patent, that as the volume of warm water at a temperature above 7° C., increases to the northwards, so the mass of cold water at a temperature below 3° C. increases towards the opening of the Atlantic into the Southern Sea.

I must now refer again to the frontispiece, and recall the general distribution of depth in the Atlantic. In discussing this question, I will speak of the eastern basin of the Atlantic, stretching from the west coast of Britain nearly to the Cape of Good Hope, and bounded to the westward by the median ridge; the north-western basin, bounded to the west and north by the coast of North America and the shoal-water extending across to Greenland, to the east by the median ridge, and to the south by the spur of the ridge joining the coast of South America at Cape Orange; and the south-western basin bounded to the north by this spur, to the west by the coast of South America, to the east by the median ridge, and to the south entirely open to the Antarctic Sea. In all our serial soundings in the eastern and the north-western basins the temperature slowly fell to a depth of about 2,000 fathoms, and from that depth it remained nearly uniform to the bottom, the difference in the readings beyond 2,000 fathoms being so slight as to be well within the limits of error of observation with Six's thermometers, but on the whole showing a tendency to sink, or at all events showing no tendency to rise on the correction for pressure being applied, which they ought to have done, had the temperature been absolutely the same. The bottom temperatures and the recorded temperatures below 2,000 fathoms were slightly but constantly lower in the north-western than they were in the eastern basin, in the former averaging about 1°·6 C., and in the latter a little under 1°·9 C.

In the south-western basin the vertical distribution of temperature is different, and this difference appears to give the key to the whole question of the distribution of temperature at great depths in the Atlantic. On our return voyage in February, 1876, four observing stations, numbered on plates XXXIV. and XXXV. from 317 to 320, were established. Two of these were in comparatively shallow water near the edge of, but still upon, the plateau which extends from the coast of South America to a distance of nearly 400 miles, and includes the Falkland Islands; the two remaining soundings, 318 and 319, were well beyond the cliff of the plateau at depths greater than 2,000 fathoms. All these soundings, the two deep ones particularly, indicate the presence of a great underlying mass of cold water, the isothermobath of 2° C. occurring at station 318 at a depth of 125 fathoms. At station 319 the 2° C. line is at 1,100 fathoms, and the other isothermobaths up to 5° C. show a corresponding rise. I attribute this remarkable difference between two soundings so near one another to the banking of the cold water against the submarine cliff by the Brazil current; sounding 318 seems to have fallen directly upon the ' cold wall.'

At the deeper sounding (319) the thermometer fell, for the first time in our experience in the South

Atlantic, below the freezing-point ; but the relations of this very low bottom-temperature will be better understood when we consider the section between Monte Video and Tristan d'Acunha.

On the line between Monte Video and Station 335, fifteen observing-stations were established. The first three of these, 321 to 323, were on the estuary of the River Plate, or (323) just beyond the edge of the delta at its mouth ; the next seven, 324 to 330, gave a section of a wide inlet into the western trough of the South Atlantic with a mean depth of 2,750 fathoms ; and the remaining five stations, 331 to 335, were on the central rise, with an average depth of 1,850 fathoms. The mean bottom-temperature of the seven deep soundings is — 0°·4 C., and that of the five soundings on the rise + 1°·3 C. ; the isothermobath of 0°·0 C. is at a depth averaging 2,400 fathoms, a depth which it never much exceeds except where the cold water rises against the American coast, as at stations 319 and 323 ; it therefore occurs in the line of the seven deep soundings only ; and there it forms the upper limit of a mass of water with a temperature below zero, 320 square miles in section. Perhaps the isothermobath of 1°·5 C. may fairly be taken as the upper limit of the very cold water; the section of the Antarctic indraught below that temperature is here about 800 square miles. (The transverse section of the Gulf-stream is about 6 square miles ; there is no volume of water at all in the Labrador current below 1°·5 C. opposite Halifax, that temperature being only found at the bottom.)

The isothermobaths of 2°, 2°·5, 3°, and 4° C. are very constant at 9,500, 900, 600, and 400 fathoms respectively, for all the stations on the parallel except station 323 on the 'cold wall,' where all the lower temperature-lines are at a much higher level, and at the shallow sounding at station 331, where all the lines below that of 4° C. rise slightly. We must be careful, however, not to attach too much importance to slight deviations of the colder lines. On the scale used in the plates, the mean interval between the isothermobaths of 2° and 3° C. in the Atlantic is 1,000 fathoms; so that a rise or fall of 100 fathoms, which is very prominent on such diagrams, actually represents only one-tenth of a centigrade degree, an amount very small in itself, and quite within the limit of error of observation with a deep-sea thermometer; it is only where there is a concordance among several isothermobathic lines in such a rise or fall that the indication is of any real value.

From these observations we learn that along the line where the south-western trough of the Atlantic joins the Southern Sea

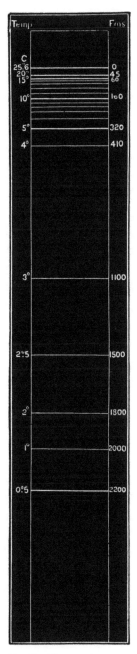

Fig. 54.—Diagram of the vertical distribution of Temperature at Station 112.

the temperature falls steadily and perceptibly to the bottom, and that the bottom temperature is more than 2° C. lower than the temperature at similar depths in the eastern or the north-western basin. The conditions which exist at the mouth of the trough extend to the equator.

Figure 54 represents the vertical distribution of temperature at station 112, lat. 3° 33′ S. long. 32° 16′ W., twenty-one miles to the north-west of Fernando Noronha. Figure 55 gives the temperature at station 129, lat. 20° 12′ S., long. 35° 19′ W., nearly midway between station 112 and station 327, one of the most characteristic in the section at present under consideration, represented in Figure 56. The depth at station 327 is 2,900 fathoms, and the depths at the two other stations 2,150 and 2,200 respectively; and it will be seen that at the latter stations the bottom-temperatures correspond almost precisely with the temperature at station 327 at like depths. The isothermobath of 2° C. is at the same height, 1,500 fathoms, at the two southern stations; and at the northern station only, near the equator, it sinks to 1,800 fathoms. The isothermobaths of 2°·5 and 3° C. correspond within a hundred fathoms or so in level at stations 129 and 327; at station 112 all the isothermobathic lines under that of 4° C. down to the line of 1° C. are much lower than at stations 129 and 327; that is to say, that at the equator, between 410 fathoms and 2,000 fathoms, the water is considerably warmer than it is further south.

The isothermobathic lines of 4° and 5° C. seem everywhere in the Atlantic to mark broadly the line of demarcation between the upper zone, where the

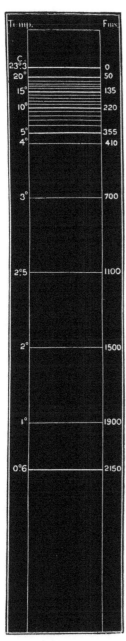

Fig. 55.—Diagram of the vertical distribution of Temperature at Station 129.

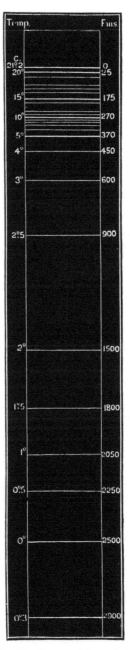

Fig. 56.—Diagram of the vertical distribution of Temperature at Station 327.

temperatures are obviously affected by the diffusion of water by wind-currents; and the lower zone, where the temperatures are continuous with those of the Southern Sea. In the North Atlantic they are markedly lower than they are to the south of the equator; that is to say, there is a much larger body of water above them heated by conduction, convection, and mixture.

The section between Monte Video and the meridian of Tristan d'Acunha includes, besides the soundings on the South-American plateau and the soundings on the 'cold wall,' a series of soundings crossing the south-western trough with an average depth of 2,750 fathoms and an average bottom-temperature of $-0°\,4$ C., and a few soundings on the middle ridge of the Atlantic, with an average depth of 1850 fathoms and a mean bottom-temperature of $+1°\!\cdot\!3$ C. There seems to be little doubt that in the trough a huge mass of Antarctic water, at temperatures ranging from $+1°\!\cdot\!5$ C. to $-0°\!\cdot\!6$ C., is creeping northwards at depths greater than 1,800 fathoms; on the central rise very little water at a temperature lower than $+1°\!\cdot\!5$ C. passes northward; but that is only on account of the absence of the required depth, for the isothermobaths of $1°\!\cdot\!5$ and $2°$ C. are practically at the same levels respectively over the central plateau and over the trough. But the evidence seems equally cogent that the water at depths less than 1,800 fathoms, and at temperatures higher than $1°\!\cdot\!5$ C. is part of the same mass, and is moving in the same direction; we can trace the same strata continuously over the trough and over the eastern and north-western basins, the temperature of each layer only very slightly rising, as has been already shown, to the northward.

Suppose a mass of water at a temperature gradually sinking from the surface downwards (Fig. 57) to be flowing slowly in a certain direction, and suppose the course of that water to be intercepted by a barrier which rises to the height of the layer of water at a temperature of 2° 0 C. Suppose at the same time that the water beyond the barrier is not constitutionally prone to alter its temperature, and that it is quietly drawn off before it has

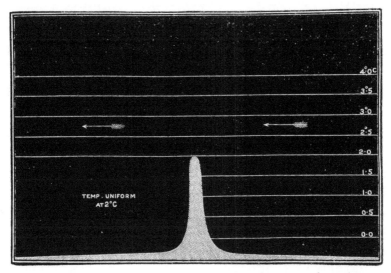

FIG. 57.—Diagram showing the effect of a 'continuous barrier' on ocean temperature.

time to do so from any external cause; it seems clear that the water beyond the barrier will be of the uniform temperature to the bottom, of the stratum of water which is passing over the barrier;—or very nearly so, for if there be any appreciable *vis à tergo*, a little water at a slightly lower temperature will force itself over the barrier and sink to the bottom.

Now, if we admit that the water in the basin of

the Atlantic consists of a continuous indraught welling into it from some cause from the Southern Sea, the southern water is welling into a space honeycombed by such barriers. On the eastern side it meets with a barrier not far to the north of the Cape of Good Hope, uniting the coast of Africa with the central ridge, and no water can pass into the eastern basin at any lower level than the lowest part of that barrier and of that ridge. On the western side of the central ridge the water passes freely up in the South-western basin nearly as far as the equator; but opposite British Guiana it is met by the barrier uniting the coast of South America with the central ridge, so that here again the ingress of all water below a certain temperature is stopped, and although the extreme depth of the North-western basin is at least 3,875 fathoms, the temperature of 1°·6 C. is maintained from a depth of 2,000 fathoms to the bottom.

All the facts of temperature-distribution in the Atlantic appear to favour the view that the entire mass of Atlantic water is supplied by an indraught from the Southern Sea, moving slowly northwards, and interrupted at different heights by the continuous barriers which limit its different basins; but this involves the remarkable phenomenon of a vast body of water constantly flowing into a *cul de sac* from which there is no exit. When I suggested this view some years ago I was asked, very naturally, how it was possible that more water could flow into the Atlantic than flowed out of it, and at that time I could see no answer to the question, although I felt sure that a solution must come some day. Now it seems simple enough; but in order to understand the conditions fully, I would ask my readers to recall the appearance of the Atlantic—and of the Pacific also, which is under exactly the same conditions— not on a map on Mercator's projection, where the northern and southern portions are necessarily greatly distorted, but on a terrestrial globe, or on such a representation of part of a globe as we have in the frontispiece to this volume. The earth may be divided into two halves, aptly called by Sir Charles Lyell the land- and the water-hemispheres, one of which contains the greater part of the ocean, while the other includes almost all the land with the exception of Australia. On the globe one sees much more clearly than on a map that the Atlantic is a mere tongue as it were of the great ocean of the water hemisphere stretching up into the land. The Arctic Ocean, with which it is in connection, is again a very limited sea, and nearly land-locked. The North Pacific is another gulf from this 'water-hemisphere,' but one vastly wider and of greater extent; while the South Pacific is included within the 'water-hemisphere.'

Although from the meridional extension of the continents to the southward, the water of the Atlantic is, as I have shown, directly continuous, layer for layer, with the water of the Antarctic basin, it must be looked upon not as being in connection with that basin only, but as being a portion of the great ocean of the water-hemisphere; and over the central part of the water hemisphere precipitation is certainly greatly in excess of evaporation, while the reverse is the case in its extensions to the northward.

The water is therefore carried off by evaporation
from the northern portions of the Atlantic and
of the Pacific, and the vapour is hurried down
towards the great zone of low barometric pressure
in the southern hemisphere; the heavy, cold water
welling up from the southward into the deepest
parts of the northward extending troughs to
which it has free access, to replace it. It is unfortunate that we have as yet scarcely sufficient data
to estimate the relative amount of rain and snow
in the northern and southern hemispheres, but the
broad fact that there is very much more in the
southern is so patent as scarcely to require proof.
This excess becomes still more apparent when we
include, as we must do, in this source of supply of
water to the north, the tropical region of the South
Pacific, which forms part of the great ocean.

To recapitulate briefly the general facts and conclusions with regard to the distribution of ocean
temperature in the Atlantic, it seems to me :—

1. That the Atlantic must be regarded in the light
of an inlet or gulf of the general ocean of the water-
hemisphere, opening directly from the Southern Sea.

2. That the water of the Southern Sea simply
wells up into the Atlantic, and that all the temperature bands of the Atlantic are essentially continuous
with like temperature bands in the Southern Sea,
with these modifications :—That (*a*) above a certain
line, which may be roughly represented by the isothermobathic lines of 5° and 4° C., the temperature
of the water is manifestly affected by direct radia-
tion and by the very complicated effects, direct and
indirect, of wind currents; and (*b*) that the whole

mass of water gradually and uniformly rises in temperature towards the head of the gulf.

3. That water at any given temperature (below 4° C.) can only occur in the Atlantic where there is a direct communication with the belt of water at the same temperature in the Southern Sea without the intervention of any continuous barrier. (The actual result of the present arrangement of such barriers is, that however great the depth may be, no water at a temperature lower than 1°·9 C. is found in the eastern basin; none at a temperature lower than 1°·6 C. in the north-western; and none beneath the freezing-point anywhere in the Atlantic, except in the depression between the coast of South America and the central ridge, to the south of the equator.)

4. That the temperature of the Atlantic is not sensibly affected by any cold indraught from the Arctic Sea. (I purposely neglect the Labrador current and the small branch of the Spitzbergen current, for these certainly do not sensibly affect the general temperature of the North Atlantic.)

5. That although there is a considerable flow of surface-water through the influence of wind-currents from the Atlantic into the the Southern Sea, that flow is not sufficient to balance the influx into the basin of the Atlantic (the constant influx being proved by the . maintenance of a general uniformity in the course of the isothermobathic lines, and by the maintenance in all the secondary basins of the minimum temperature due to the height of their respective barriers); that for several reasons (the lower barometric pressure and the sup-posed greater amount of rainfall in the Southern

Sea; the higher specific gravity at the surface than
at greater depths in the Atlantic; the higher specific
gravity of the surface-water in the Atlantic to the
north than to the south of the equator), it is probable that the general circulation is kept up chiefly
by an excess of evaporation in the region of the
North Atlantic, balancing a corresponding excess
of precipitation over evaporation in the water-
hemisphere.

The distribution and nature of the deep-sea fauna.—
The most prominent and remarkable biological result
of the recent investigations is the final establishment of the fact that the distribution of living
beings has no depth-limit; but that animals of all
the marine invertebrate classes, and probably fishes
also, exist over the whole of the floor of the ocean:
and some of the most interesting of the problems
which are now before us have reference to the nature
and distribution of the deep-sea fauna, and to its
relations with the fauna of shallower water, and with
the faunæ of past periods in the earth's history.
This is however precisely the class of questions
which we are as yet least prepared to enter into,
for everything depends upon the careful study and
the critical determination of the animal forms which
have been procured; and this task, which will occupy
many specialists for several years, has been only just
commenced.

My present impression is that although life is thus
universally extended, the number of species and of individuals diminishes after a certain depth is reached,
and that at the same time their size usually decreases.
This latter observation is not, however, true for all

groups; a peculiar family of the Holothuridea, very widely distributed in deep water, maintain the full dimensions of the largest of their class, and even exhibit some forms of unusual size. Of the value of our present impressions on such questions I am by no means sure. Using all precautions, and with ample power and the most complete appliances, it is extremely difficult to work either with the dredge or with the trawl at depths approaching or exceeding 3,000 fathoms. A single dredging operation in such depths takes a long time; the dredge is put over at daybreak, and it is usually dark before it is recovered, so that the number of such operations must be comparatively small. It is necessary to take every precaution to keep the ship as nearly as possible in the same place; and as this can never be done absolutely, it is unsafe to run the risk of adding to any motion which the dredge may already have acquired, by attempting to drag it for any distance over the ground. The consequence is, that in those cases where the dredge does reach the bottom, it probably too often sinks at once into the soft ooze and remains clogged with a single 'mouthful' until it is hauled up again. Sometimes a slight excess of movement in the vessel, from currents or from wind-drift, seems to give a vibratory motion to the enormous length of rope, and to keep the dredge tripping over the ground, so that only a few things are picked up by the tangles or clinging to the outside of the net. We must therefore bear in mind that only an infinitesimally small portion of the floor of the ocean at depths over 2,500 fathoms has yet been explored.

Whatever may be the case at the extreme depths referred to, there can be no doubt that at depths which may be regarded as comparatively accessible, say a little above 2,000 fathoms, the fauna is sufficiently varied. I give in Appendix B. to this chapter a Table taken from the 'Station-book,' showing the number of occurrences of representatives of the principal groups of marine animals at the fifty-two stations at which we dredged or trawled successfully at depths greater than 2,000 fathoms during the voyage. All the groups marked with an asterisk on this list were represented, having been observed and noted when the trawl or dredge came up; it is very probable that on going over the collection carefully it will be found that many, particularly of the smaller forms, have been omitted. The occurrences of fishes, of cephalopods, and of decapod crustaceans must be taken with a reservation; for it is not always possible to determine whether they were taken on the bottom, or above it during the hauling in of the net.

The distribution of life evidently depends in a marked degree either upon the nature of the bottom or upon the conditions which modify the nature of the bottom. Thus over the vast areas where the calcareous matter of the pelagic foraminifera has been removed, and the bottom consists of red or grey clay, animal life is scarce; and is represented chiefly by shell-less orders, such as the Holothuridea and the Annelids. This comparative sterility depends no doubt to a great degree upon the absence of carbonate of lime, but not entirely so; for the most sterile regions of the whole sea are the mortar-like

PLATE XLI. *Meteorological Obs*

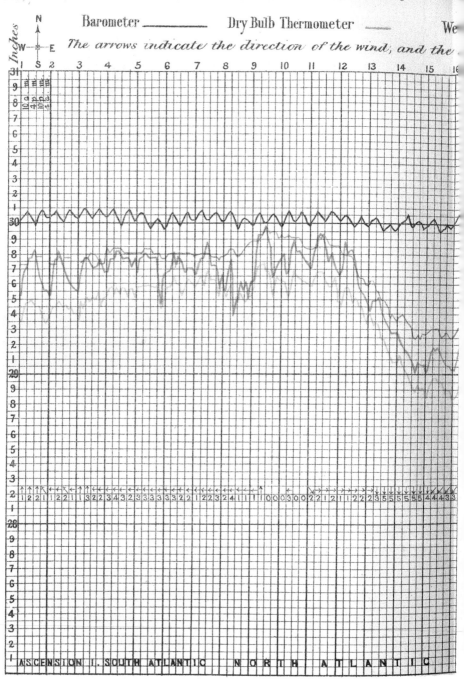

rvations for the month of April, 1876.

Bulb Thermometer ———   Temperature of Sea Surface ————

numbers beneath its force according to Beaufort's scale

17 18 19 20 21 22 23 24 25 26 27 28 29 30   40 °C

PORTO
PRAYA   PORTO GRANDE ST VINCENT

lime deposits which form the slopes of coral reefs and islands. There appears to be something in the state of aggregation of the lime in the *Globigerina* shells and its intimate union with organic matter which renders the globigerina-ooze a medium peculiarly favourable to the development of the higher forms of life ; the stomachs of the more highly organized animals living in it or on its surface are always full of the fresher foraminiferal shells, from which they undoubtedly derive not only material for the calcification of their tests, but nitrogenous matter for assimilation likewise.

As we had previously anticipated, the fauna at great depths was found to be remarkably uniform. Species nearly allied to those found in shallow water of m ny familiar genera were taken in the deepest hauls, so that it would seem that the enormous pressure, the utter darkness, and the differences in the chemical and physical conditions of the water and in the proportions of its contained gases depending upon such extreme conditions, do not influence animal life to any great extent.

The geographical extension of any animal species, whether on land or in the sea, appears to depend mainly upon the maintenance of a tolerably uniform temperature, and the presence of an adequate supply of suitable food, the latter condition again depending chiefly upon the former; and the conditions both of temperature and of food-supply are very uniform at extreme depths where the nature of the bottom is the same. Possibly the element next in importance is the length of time during which migration may have taken place, and there seems much reason for believing that the great ocean-depressions of the present time have persisted through all the later geological periods, back probably as far as the Permian age, and perhaps much farther. If this be so, the length of time during which the vast area occupied by the abyssal fauna has maintained its continuity, and probably a great uniformity in essential conditions, is incalculable; that is to say, it cannot in the present state of our knowledge be reduced even approximately to astronomical time.

In discussing the general distribution of temperature the reasons have been already given which have led us to the belief that there is a constant under-flow of water from the south northwards, and one would naturally expect some indication of migration having proceeded, and continuing to proceed in that direction. It is impossible to come to a definite conclusion on this question until the species in the different groups shall have been critically determined; there seems however to be little doubt that the families which are specially characteristic of the abyssal fauna, such as the Hexactinellid sponges, the stalked Crinoids, the Echinothuridæ, and the genera allied to *Infulaster* and *Micraster* among the Echinidea, are more abundant, and larger and more fully developed, in the Antarctic ocean and in the great ocean of the water-hemisphere generally, than they are in the Atlantic and the North Pacific.

Our preliminary dredgings in the North Atlantic along the coasts of Portugal and Spain were chiefly on the globigerina-ooze at depths under 2,000 fathoms; and there we found all the ordinary forms of deep-sea life abundant, particularly sponges referable to the genera *Hyalonema, Aphrocallistes, Euplectella, Coral-listes*, and *Caminus*.   As this area had been gone over by Mr. Gwyn Jeffreys in the 'Porcupine' we were already aware that stalked crinoids and corals of Tertiary types occurred.

The first section across the Atlantic, from Teneriffe to Sombrero, was through deep water, and princi-pally over a bottom of red clay, the most unproductive of all the deep-sea sediments.   The following Table gives an idea of the proportion in which the principal zoological groups were represented :—

	Station 1. 1890 fms.	Station 2. 1945 fms.	Station 3. 1530 fms.	Station 5. 2740 fms.	Station 9. 3150 fms.	Station 13. 1900 fms.	Station 14. 1950 fms.	Station 20. 2975 fms.	Station 22. 1420 fms.	Station 23. 450 fms.
Pisces	*	...	...	...	...	...	...	...	*	*
Cephalopoda	*									
Gastropoda	*	...	...	...	...	...	*		...	*
Lamellibranchiata . . .	...	...	...	*	...	...	*	...	...	*
Brachiopoda	...	...	...	...	...	...	*			
Tunicata	*									
Decapoda	...	...	...	...	...	...	*			*
Schizopoda	*									
Edriophthalmata . . .	*	...	*							
Copepoda	*									
Annelida	*	...	*	...	...	...	...	*	...	*
Gephyrea	...	*	...	...	...	...	...	...	...	*
Bryozoa	...	...	...	*	...	...	*	*	...	*
Echinoidea	...	...	..	...	...	...	...	...	...	*
Ophiuridea	...	...	*	...	...	...	...	...	...	*
Asteridea	...	...	*	...	...	...	...	...	...	*
Hydromedusæ	*	...	*	...	...	...	...	...	...	*
Alcyonaria	...	...	*	...	...	...	...	...	...	*
Porifera	...	...	*	...	...	*	*	...	...	*

The only stations in this section which can be considered at all productive are No. 3 and No. 13, both on globigerina-ooze, and Station 23 in shallow water off the Island of St. Thomas.   At the other stations animal forms were few in number and appa-rently stunted in growth.

In the next series of stations, from Bermudas to Sandy Hook and Halifax and back to Bermudas, the conditions varied greatly; but by far the greatest abundance of animal life occurred in the comparatively shallow water, including one or two of the cod banks off the American coast and the coast of Nova Scotia. The fauna of that region was of course on the whole well known; some interesting observations were however made on the distribution of the sub-arctic fauna in deeper water. At one or two stations off the edge of the banks several species of the curious *Infulaster*-like genus *Pourtalesia* occurred, but extremely small and dwarfed, a great contrast to the fully-developed forms of the same group which are abundant in the Antarctic sea.

	Station 24. 390 fms.	Station 29. 2700 fms.	Station 33. 435 fms.	Station 36. 32 fms.	Station 40. 2675 fms.	Station 44. 1700 fms.	Station 45. 1250 fms.	Station 46 1350 fms.	Station 47. 1340 fms.	Station 48. 51 fms.	Station 49. 83 fms.	Station 50. 1250 fms.	Station 54. 2650 fms.	Station 56. 1075 fms.	Station 57. 690 fms.
Pisces					*										
Cephalopoda											*				
Gastropoda														*	
Lamellibranchiata				*				*		*		*		*	
Brachiopoda	*									*	*	*			
Tunicata	*					*			*	*					
Pycnogonida												*	*		
Decapoda	*	*	*	*	*	*	*			*	*	*	*	*	*
Stomatopoda							*								
Edriophthalmata												*	*		
Copepoda								*							
Cirripedia	*														
Annelida	*		*	*		*		*		*		*	*	*	
Gephyrea	*			*		*	*			*		*	*		
Bryozoa											*				
Holothuridea	*							*				*	*		
Echinoidea			*			*	*			*	*			*	
Ophiuridea	*		*			*	*	*	*	*		*	*	*	
Asteridea	*			*		*	*	*	*	*		*	*		
Crinoidea	*										*				
Hydromedusæ				*							*				
Zoantharia	*			*				*	*			*	*		*
Alcyonaria	*		*			*		*	*			*		*	*
Porifera	*		*	*											*

Although most of the dredgings between Bermudas and Madeira, with the exception of a few near the Açores, were in very deep water, animal life was fairly represented; and some groups, the Cirripedia for example, yielded one or two of their largest and most striking species.

	Station 61. 2850 fms.	Station 63. 2750 fms.	Station 64. 2750 fms.	Station 68. 2175 fms.	Station 69. 2200 fms.	Station 70. 1675 fms.	Station 71. 1675 fms.	Station 72. 1240 fms.	Station 73. 1000 fms.	Station 76. 900 fms.	Station 78. 1000 fms.	Station 79. 2025 fms.	Station 83. 1650 fms.
Pisces	*												
Gastropoda						*				*		*	
Lamellibranchiata						*	*	*					
Brachiopoda										*			
Tunicata						*							
Pycnogonida						*							
Decapoda		*			*						*	*	
Schizopoda		*			*								
Edriopthalmata				*		*					*	*	
Cirripedia	*	*									*		
Annelida	*	*	*			*					*	*	*
Gephyrea			*										
Polyzoa				*	*		*				*	*	
Holothuridea	*												*
Echinoidea						*				*			*
Ophiuridea	*			*							*		*
Asteridea						*					*	*	*
Crinoidea											*		
Hydromedusæ		*									*		
Zoantharia		*		*		*					*		*
Alcyonaria						*						*	
Porifera	*			*		*	*				*	*	

The six stations on the section between Madeira and Station 102 were mostly in water of moderate depth on a line parallel with the coast of Africa, and sufficiently near the coast to have the deposits sensibly influenced by the presence of land detritus. Such an admixture of river or shore mud is usually unfavourable to the development of a rich fauna, and the number of groups represented is accordingly small.

	Station 85. 1125 fms.	Station 87. 1675 fms.	Station 89. 2400 fms.	Station 92. 1975 fms.	Station 98. 1750 fms.	Station 101. 2500 fms.
Pisces	...	...	*	*	...	*
Gastropoda	...	...	*			
Lamellibranchiata . . .	*	...	...	...	*	
Decapoda	*	*	...	...	...	*
Schizopoda	...	...	...	*		
Edriophthalmata . . .	...	...	...	...	...	*
Copepoda	...	...	*			
Cirripedia	...	...	*			
Annelida	*	...	...	...	...	*
Gephyrea	...	...	...	...	...	*
Polyzoa	*	*	*	...	...	*
Holothuridea	...	...	*	*	...	*
Echinoidea	...	...	...	...	*	
Ophiuridea	*	*	*			
Asteridea	...	...	*			
Alcyonaria	*	*	*	...	...	*
Porifera	*	...	...	...	*	*

Of the next series of stations where the trawl or dredge was employed successfully, the first three, 104, 106, and 107, were in deep water nearly under the line; Station 109 was in shallow water near St. Paul's rocks, Station 103 A was close to the island of Fernando Noronha, and the remainder were at moderate depths, usually much below 1,000 fathoms, along the Brazilian coast from Cape San Roque southwards to Bahia. The fauna of course varied greatly in this section with the varying conditions. Along the coast of Brazil the bottom was usually river-mud more or less mixed with the shells of globigerina and the débris of surface shells; and the fauna was comparatively rich, recalling that of the western coast of South Europe in the abundance of hexactinellid and coralloid sponges.

	Station 104. 2500 fms.	Station 106. 1850 fms.	Station 107. 1500 fms.	Station 109. 100 fms.	Station 113 A. 7—25 fms.	Stns. 120 and 121. 675 and 500 fms.	Station 122. 350 fms.	Station 122. 30 fms.	Station 122. 400 fms.	Station 124. 1600 fms.	Stns. 125 and 126. 1200 and 770 fms.
Pisces	*	*	*	...	*	*	*	*	...	*	*
Cephalopoda	...	...	...	...	...	...	...	*			
Gastropoda	*	*	...	...	*	...	*	...	*	*	
Lamellibranchiata	*	*	...	*	*	*	*	*			
Brachiopoda	...	*	...	...	...	...	*				
Decapoda	*	*	*	*	*	*	*	*	*	...	*
Schizopoda	*	*	*	...	...	*	*	...	...	...	*
Edriophthalmata	...	...	...	...	*						
Cirripedia	*										
Annelida	*	*	...	...	*	*	*	*	...	*	*
Gephyrea	...	...	...	*							
Polyzoa	*	*	...	*	...	...	*	...	*		
Holothuridea	*	...	...	...	...	...	*	...	...	...	*
Echinoidea	...	*	...	*	*	*	*	*	...	*	
Ophiuridea	*	*	...	*	*	*	*	*	*		
Asteridea	...	*	...	*	...	*	*	*	...	...	*
Crinoidea	...	*	...	*	...	*	*	...	*		
Hydromedusæ	...	...	...	...	...	*					
Zoantharia	...	...	...	*	*	*	*	*	*		
Alcyonaria	...	*	...	*	*	...	*	*	...	*	
Porifera	...	...	*	...	...	...	*	*	...	*	

The following Table gives the general distribution of the principal animal groups along a line extending from the coast of South America to the Cape of Good Hope, nearly along the parallel of 40° south. Most of these dredgings were in comparatively deep water, some on the grey and red clays of the western and eastern troughs, and several on the median ridge of the Atlantic. Along this line, which may be said to indicate the limit between the Atlantic and the Southern sea, the forms which are specially abyssal, and which are most nearly related to extinct chalk or older tertiary species, are certainly more fully developed and more numerous than they are in any part of the Atlantic ' gulf.'

	Station 131. 2275 fms.	Station 133. 1900 fms.	Station 134. 100—150 fms.	Station 135. 1000 fms.	Station 137. 2550 fms.	Station 322. 21 fms.	Station 323. 1900 fms.	Station 325. 2650 fms.	Station 331. 1715 fms.	Station 332. 2200 fms.	Station 333. 2025 fms.	Station 334. 1915 fms.	Station 335. 1425 fms.
Pisces		*			*		*	*		*			*
Cephalopoda											*		
Gastropoda			*				*	*					*
Lamellibranchiata		*			*	*		*					*
Tunicata		*											
Pycnogonida													*
Decapoda		*		*	*		*	*	*				*
Schizopoda		*			*			*			*		
Edriophthalmata												*	
Cirripedia				*									
Annelida							*	*	*			*	*
Polyzoa							*	*			*	*	
Holothuridea			*				*	*					
Echinidea				*	*								*
Ophiuridea				*			*	*			*	*	
Asteridea		*	*		*		*	*					*
Crinoidea				*			*						
Hodromedusæ							*						
Zoantharia	*	*	*	*			*	*	*	*			
Alcyonaria		*	*	*								*	
Porifera			*	*			*	*			*	*	*

It may not be out of place, before leaving this subject, to give a brief preliminary sketch of the distribution of the groups of marine organisms which inhabit the depths of the sea; or, leading a pelagic existence, contribute by the subsidence of their hard-parts after death, to the formation of sub-marine deposits. This is a subject which must be much more fully discussed when the species have been determined, and the new forms described, but we have already perhaps sufficient material for a general outline.

No plants live, so far as we know, at great depths in the sea; and it is in all probability essentially inconsistent with their nature and mode of nutrition that they should do so. What may be their extreme limit I am not prepared to say; some straggling plants may occur at much greater depths, but certainly what is usually understood by *vegetation* is practically limited to depths under 100 fathoms. Very few of the higher Algæ live even occasionally on the surface of the sea; the notable exception is the gulf-weed (*Sargassum bacciferum*), which scatters its feathery islets over vast areas of warm, still water; and affords rest and shelter to the peculiar nomadic fauna to which I have already alluded (vol. i. p. 186, &c.).

Confervoids and unicellular Algæ occur, however, frequently, and sometimes in such profusion as to discolour the water over an area of many miles. If Diatoms are to be regarded as plants, these are found abundantly on the surface, more particularly where the specific gravity of the water is comparatively low. The frustules of Diatoms occur in all the deep-sea deposits in greater or less number; and in some places, as at a few of the stations in the Indian Ocean, they form the bulk of the sample brought up by the sounding machine. Over the area occupied by this siliceous deposit, the higher fauna were found to consist mainly of forms with but little carbonate of lime entering into the composition of their tests, such as very thin-shelled irregular urchins, and especially an abundance of Holothuridea. These were often modified in a singular way; the perisom was reduced to a mere membrane, and the stomach and intestine were expanded so as to occupy nearly the whole of the body cavity; and distended with the 'diatom-ooze,' so completely, that the animal looked like a thin transparent bag filled with it. There can be little doubt that the diatoms sink to the bottom still retaining a small portion of their organic matter,

z 2

which is slowly extracted by the alimentary canal of the Holothurid.

Radiolarians were met with throughout the whole of the Atlantic; and often in great abundance, the sea being not unfrequently slightly discoloured by them. The forms which occurred in such numbers were usually species of the Acanthometridæ, but Polycystina and the compound genera were also numerous. The remains of Radiolarians were found in all deep-sea deposits, usually in very direct proportion to the numbers occurring on the surface and in intermediate water. It was frequently observed, however, that where, in deep water, certain species swarmed on the surface, very few of their skeletons could be detected on the bottom. This applies especially to the Acanthometridæ, and is probably owing to the extreme tenuity of the siliceous wall of their radiating spicules, which may admit of their being dissolved while sinking to a great depth; or possibly the spicules may never become thoroughly silicified, but may retain permanently more or less the condition of acanthin. The Polycystina seem much less destructible, and occur in abundance on the bottom at the greatest depths. Although the Radiolaria are universally distributed—like the Diatoms, but in a less marked degree, they seem to be most numerous where the specific gravity of the water is low; they specially swarm in the warm and comparatively still region of the South Western Pacific and among the Islands of the Malay Archipelago, where they are much more abundant than in any part of the Atlantic; I have already given the reasons which led us to the belief that Radiolarians inhabit the water of the ocean throughout its entire depth, or, at all events, its upper and lower portions.

In the investigations with the towing-net, made by Mr. Murray during the latter part of the cruise—at all depths, the nets being either sent down independently to the depths required, or attached to the dredge or trawl-rope—about thirty species or more were procured of a beautiful group of minute forms approaching, but in many important points differing from, the Radiolarians. This order have apparently hitherto escaped observation, and I retain for the type genus the name *Challengeria*, and for the Order that of 'Challengerida.' This appears to be the only new group of higher than generic value which has come to light during the 'Challenger' Expedition.

As a rule these forms are extremely minute, although some of them approach in size the smaller Radiolarians. They consist usually of a single chamber of silica, varying greatly in form, sometimes triangular, sometimes lenticular, and frequently nearly globular or flask-shaped; with a single opening, usually guarded by a beautifully formed and frequently highly-ornamented lip. The surface of the shell is usually richly sculptured, a favourite style of ornament being a series of closely apposed and sym-

Fig. 58. — *Challengeria*. Four hundred times the natural size.

metrically arranged circular pits sunk deep in the siliceous wall, their inner walls refracting the light and giving the surface of the whole a peculiar pearly

lustre.   The contents of the shell consist of a mass of
granular sarcode, with one or more large well-defined
granular nuclei, which colour deeply with carmine;
and a number of dark-brown, sometimes nearly black,
rounded compound granular masses.   It is singular
that these deeply pigmented spheres, which probably
represent the spheres of a lighter colour which we
find in all the surface rhizopods, seem to be specially
characteristic of rhizopods from deep water, being

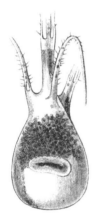 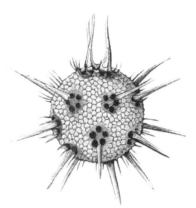

Fig. 59.—Forms of the *Challengerida.*

found also in the Radiolarians from the deep tow-
nets.   The Challengerida were never met with on the
surface, they were taken rarely in tow-nets sunk to
depths of 300 and 400 fathoms, and they were most
abundant when the tow-nets were sent down on the
dredge or trawl-rope to much greater depths.   Their
distribution seems to have a wide extension; they
are occasionally found in the bottom deposits, but
rarely, probably on account of their small size and
the extreme tenuity of their tests, which renders

them liable to solution in sea-water. The Challenge-rida are essentially rhizopods with monothalamous siliceous shells; and their zoological position may be not very far from such forms as *Gromia*.

The distribution of the pelagic Foraminifera has already been discussed. They are universally distributed throughout the temperate and warmer seas, diminishing in number and decreasing in size towards the frigid zones. Certain species are occasionally found in large numbers on the surface, but at a depth of a few fathoms their occurrence is much more certain. We have good reason to believe that the vertical range of the oceanic group does not extend beyond the first few hundred fathoms, and that all the pelagic forms occur occasionally on the surface. Living Foraminifera are very generally distributed on the bottom, but the forms differ from those found on the surface and near it, and are for the most part to be referred to arenaceous or imperforate types.

Sponges extend to all depths, but perhaps the class attains its maximum development between 500 and 1,000 fathoms. All the orders occur in the abyssal zone, except the Calcarea, which seem to be confined to shallow water. At great depths the Hexactinellidæ certainly preponderate; and next to these perhaps the Esperiadæ, the Geodidæ, and the Lithistidæ; the ordinary horny and halichondroid forms, although they have a considerable vertical range, are most abundant in the coralline zone. In the Atlantic, hexactinellid sponges are very abundant to depths of about a thousand fathoms along the coasts of Portugal and Brazil; these forms, which occur in the fossil state in the earlier palæozoic rocks, and, represented by the

Ventriculidæ and allied families abound in the chalk and greensand, show in a marked degree the wide extension in space at the present day of a very uniform abyssal fauna, the same or very similar species of the genera *Aphrocallistes, Farrea, Hyalonema, Euplectella, Holtenia,* and *Rossella* being apparently cosmopolite. Nearly all the deep-sea sponges of all orders are stalked, or provided with beards or fringes of radiating spicules, or otherwise supplied with means of supporting themselves above the surface of the soft ooze in which they grow.

Among the Cœlenterata the Hydrozoa are not very fully represented at great depths. To this rule, however, some singular exceptions occur. In many of our deepest dredgings, where there was a great lack of carbonate of lime, and animal life appeared to be very scarce, the curved horny tubes of what is probably a species of the genus *Stephanoscyphus* was found adhering to the ear-bones of whales or to concretions of iron and manganese; and on two occasions in the North Pacific, at depths of 1,875 and 2,900 fathoms, we captured a giant of the class, a species of *Monocaulus* with a stem upwards of two metres long and a head three or four decimetres across the crown of extended tentacles.

True corals referable to the Madreporaria are not abundant in deep water. According to Mr. Moseley's report about ten genera reach a depth of 1,000 fathoms; four genera are found at 1,500 fathoms, and a single species extends practically through all depths, ranging from 30 to 2,900 fathoms. In the Atlantic especially deep-sea corals are sparsely scattered; two or three species of the genus *Caryophyllia* are among the most common, and *Deltocyathus Agassizii*, and one or two species of the genus *Ceratotrochus*, were frequently met with near the American

FIG. 60.—*Flabellum apertum*, MOSELEY. Natural size.

coast and in the Gulf-stream region. Besides *Flabellum alabastrum*—the fine species already described from the Açores—*F. apertum* (Fig. 60), a form with a wide geographical range, occurred off the coast of Portugal; and a very delicate little species,·named by Mr. Moseley *Flabellum angulare* (Fig. 61), was dredged on one occasion only, not far from the fishing banks of Nova Scotia, at a depth of 1,250 fathoms. The special peculiarity of this species, if the individual which we procured be not abnormal, is its regularly pentagonal form and the perfect quinary arrangement of its parts; it

FIG. 61.—*Flabellum angulare*, MOSELEY. Natural size.

has exactly 40 septa—10 primary and secondary, 10 tertiary, and 20 quaternary. Species of *Lopho-*

helia and of *Amphihelia* were generally distributed at comparatively moderate depths, and the cosmopolitan *Fungia symmetrica* occurred in small number at all depths. The deep-sea corals are mostly simple and solitary, and the greater number belong to the Turbinolidæ; nearly all the genera pass back to tertiary, and a few to mezozoic times. Upon the whole the corals must undoubtedly be regarded as affording evidence of a certain relation between the deep-sea fauna of the present day and the fauna of shallower water during the deposition of at all events some portions of the tertiary series.

Attached Alcyonarians, and especially genera allied to *Mopsea* and *Primnoa*, are extremely abundant in the cooler seas at depths from 500 to 1,000 fathoms, sometimes occurring in such quantity as to hamper and clog the trawl, and affording charming exhibitions of elegance of form and beauty of colouring. Certain forms of the Pennatulidæ go down to great depths; the genus *Umbellula*, which we at first regarded as of extreme rarity, turned up every now and then, usually in nearly the deepest hauls, represented by two or three nearly allied species.

Among the Echinodermata the stalked crinoids of the deep-sea fauna are most interesting, but they are comparatively few in number. The large forms belonging to the Pentacrinidæ, although they are very local, appear to be more common than has been hitherto supposed at depths of from three to five hundred fathoms. Five or six new species have been added to the meagre list, but most of these are from the South-western Pacific and do not enter into the Atlantic fauna. The Apiocrinidæ, represented by the genera *Rhizocrinus*, *Bathycrinus*, and *Hyocrinus*, which are of so great interest as the last survivors of a large and important order, are rare prizes at much greater depths. Representatives of all the three genera were dredged in deep water in the South Atlantic.

Ophiuridea, many of them referable with the common sand brittle-star to the genus *Ophioglypha*, and many others to the closely-allied genus *Ophiomusium*, came up from the greatest depths, and, particularly in the North Atlantic, formed a prominent feature in the fauna. Asteridea, principally represented by forms more or less nearly allied to *Astropecten*, *Astrogonium*, *Archaster*, *Pteraster*, and *Hymenaster*, abounded at all more moderate depths; and the singular aberrant genus *Brisinga* was found universally from the coast of Labrador to the Antarctic ice-barrier, at all depths, from 400 to 3,000 fathoms, the trawl rarely coming up from deep water without some fragments of its fragile arms.

The novel forms of sea-urchins, regular and irregular, are numerous and highly interesting especially in their palæontological aspect. Species of the genera *Porocidaris* and *Salenia* occur not unfrequently, and the curious flexible Echinothuridæ have assumed the proportions of an important family. Among the irregular urchins the relation between the modern abyssal fauna and the fauna of the later mezozoic beds is even more marked; a number of genera hitherto undescribed associate themselves with the chalk genus *Infulaster*, while others find their nearest allies in *Micraster* and *Ananchytes*.

The Holothuridea are very generally distributed

down to the greatest depths; and are represented in deep water by a peculiar series allied to *Psolus*, with a very distinct ambulatory disk, very frequently a great development of calcified tissue in the perisom, and frequently symmetrical series of long tubular appendages along the back and sides. These *Holothuriæ*, which are among the most characteristic of the abyssal forms, have not yet been critically examined.

Polyzoa were found at all depths; some extremely beautiful and delicate forms referred principally to the Bicellariadæ and to the Salicornariadæ occurred at depths between 2,000 and 3,000 fathoms in sterile regions where other animal life was scarce.

The Gephyrea yielded a few interesting undescribed forms. Annelids were not abundant at great depths; but on one or two occasions, as for example at Station 19 on the section between Teneriffe and Sombrero, their occurrence was of special interest, for they seemed to be almost the sole inhabitants of 'red clay' from which nearly the whole of the carbonate of lime had been removed.

The various orders of Crustacea form a most interesting and important element in the ocean-fauna. The pedunculated Cirripedia seem to be universally distributed in comparatively small numbers even at the greatest depths, where some of the abyssal species are larger and more highly ornamented than those previously known from shallow water. Some of the finest additions to our knowledge of species were made among the Schizopoda, in colossal forms of the genera *Gnathophausia* and *Petalophthalmus*.

The Macrourous Decapods were very many, and

included some splendid undescribed species, especially among the Peneid and Caridid shrimps; there was often, however, some slight doubt whether these forms lived actually on the bottom; we had good evidence that they lived near the bottom, but in several instances shrimps were captured when we had reason to suspect that the trawl had been buoyed up and had never actually touched the ground. *Galatheæ* were frequent to great depths, but Brachyourous Decapods appear to be confined almost entirely to comparatively shallow water.

The Pycnogonida occurred frequently and attained an enormous size in cold Arctic and Antarctic water at medium depths. The Brachiopoda we found widely distributed but by no means numerous either as to species or individuals; on one or two occasions in the 'Porcupine' we got fine hauls of *Terebratula cranium* and *T. septata* attached to the pebbles of a gravel of the volcanic rocks of the Faröes, and we took one or two other species with the conditions almost repeated in the neighbourhood of the Heard Islands and the Crozets in the southern sea.

The two great modern groups of the Mollusca, the Lamellibranchiata and the Gastropoda, do not enter largely into the fauna of the deep sea. Species of both groups, usually small and apparently stunted, were widely though sparsely diffused, and exceptionally a large and handsome form occurred, as for example a singularly beautiful volute in 1,600 fathoms at Station 147 in the southern sea, some fine species of *Margarita* in 1,260 and 1,675 fathoms south of Kerguelen, and a large bivalve, allied to *Lima*, which turned up in deep dredgings

at rare intervals at stations the most widely separated
in the Atlantic and the Pacific.

Cephalopods came up in the trawl occasionally,
but in most cases they belonged to the peculiar
gelatinous group which are well known to be pelagic,
and had doubtless been taken while the trawl was
passing through the upper water. In some few
cases species had evidently come from the bottom,
but not from any great depth. It is singular that
only on one occasion we took a specimen of the

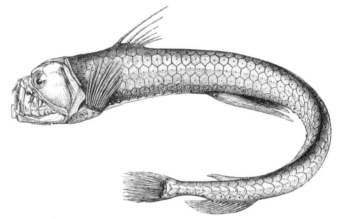

Fig. 62.—*Chauliodus s'loanii.* From the upper water. One third the natural size.

animal of *Spirula,* although the delicate little white
coiled shell is one of the commonest objects on the
beach throughout the tropics—sometimes washed up
in a long white line which can be seen from any
distance.

After the method of dredging with the trawl was
introduced one or two or more fishes were taken at
almost every haul; showing, that while not abundant
they were universally present; with these, however,
as with the decapod crustaceans, the question often

arose whether the specimen had been brought up
from the bottom, or had been taken by the trawl
on its way up. In many cases this could not be
answered with certainty, but it seems that certain
families which are met with very frequently—such
as the Sternoptychidæ and the Scopelidæ, many of
them remarkable for their grotesque forms, their
brilliant colouring, and metallic lustre, and the symmetrical rows of deeply-pigmented sense or phosphorescent organs which sometimes extend along
the greater part of the body (Fig. 62),—are in most,
if not all cases from the upper waters; while certain
other families—for example, the Ophidiidæ and the
Macruridæ—live at or near the bottom. What we
know of the distribution of fishes seems to me to
corroborate the view that in a deep-sea vertical
section there are two regions—one within a limited
distance of the surface, and the other a little way
above the bottom—which have their special faunæ;
while the zone between is destitute of at all events
the higher forms of animal life.

In some places both in the Atlantic and the Pacific,
especially at extreme depths in the red-clay areæ,
the trawl brought up many teeth of sharks and ear-
bones of whales, all in a semi-fossil state, and usually
strongly impregnated with, or their substance to a
great extent replaced by, the oxides of iron and manganese. These deposits of bones occur at great distances from land, and where from other causes the
deposition of sediment is taking place with extreme
slowness. The sharks' teeth belong principally to
genera, and often to species which we believe to be
now extinct, and which are characteristic of the later

tertiary formations; and there seems little doubt
that they have been lying there, becoming gradually
buried in the slowly-accumulating sediment, from
tertiary times. The fishes which were collected
during the expedition are now undergoing examination by Dr. Günther, and the semi-fossil remains
from the sea-bottom by Mr. Murray; and several
questions of great interest must be left open until
their investigations are completed.

The first general survey of the deep-sea collections,
undertaken with a knowledge of the circumstances
under which the specimens were procured, justify
us, I believe, in arriving at the following general
conclusions:—

1. Animal life is present on the bottom of the
ocean at all depths.

2. Animal life is not nearly so abundant at extreme, as it is at more moderate depths; but, as
well-developed members of all the marine invertebrate
classes occur at all depths, this appears to depend
more upon certain causes affecting the composition
of the bottom deposits, and of the bottom water in-
volving the supply of oxygen, and of carbonate of
lime, phosphate of lime, and other materials necessary for their development, than upon any of the
conditions immediately connected with depth.

3. There is every reason to believe that the fauna
of deep water is confined principally to two belts,
one at and near the surface, and the other on and
near the bottom; leaving an intermediate zone in
which the larger animal forms, vertebrate and invertebrate, are nearly or entirely absent.

4. Although all the principal marine invertebrate

groups are represented in the abyssal fauna, the
relative proportion in which they occur is peculiar.
Thus Mollusca in all their classes, Brachyourous
Crustacea, and Annelida, are on the whole scarce;
while Echinodermata and Porifera greatly preponderate.

5. Depths beyond 500 fathoms are inhabited
throughout the world by a fauna which presents
generally the same features throughout; deep-sea
genera have usually a cosmopolitan extension,
while species are either universally distributed, or,
if they differ in remote localities, they are markedly
representative; that is to say, they bear to one
another a close genetic relation.

6. The abyssal fauna is certainly more nearly
related than the fauna of shallower water to the
faunæ of the tertiary and secondary periods, although
this relation is not so close as we were at first
inclined to expect, and only a comparatively small
number of types supposed to have become extinct
have yet been discovered.

7. The most characteristic abyssal forms, and those
which are most nearly related to extinct types, seem
to occur in greatest abundance and of largest size
in the southern ocean; and the general character of
the faunæ of the Atlantic and of the Pacific gives
the impression that the migration of species has
taken place in a northerly direction, that is to say,
in a direction corresponding with the movement of
the cold under-current.

8. The general character of the abyssal fauna
resembles most that of the shallower water of high
northern and southern latitudes, no doubt because

the conditions of temperature, on which the distribution of animals mainly depends, are nearly similar.

The Density of Sea-water.—The specific gravity of the surface-water was determined daily by Mr. J. Y. Buchanan, the chemist to the Expedition, with great accuracy; the specific gravity of the bottom-water was also determined so far as possible at every observing station, and every opportunity was taken to procure for physical and chemical examination samples of water from intermediate depths. On our return home through the Pacific, Mr. Buchanan, at my request, prepared a preliminary report on his method of investigation and on the general results of his work, which I received at Valparaiso; and from that report the following summary of specific gravity conditions in the Atlantic, according to the first year's observations, is taken. The apparatus in use for procuring water from the bottom and from intermediate depths has been already described (vol. i. p. 34 *et seq.*).

Representing the specific gravity of distilled water at 4° C. by 100,000, Mr. Buchanan found that of ocean-water at 15°·56 C. to vary between the extremes of 102780 and 102400; so that, to be of any value at all, the possible error in the results must not exceed 10. The hydrometer used for these observations is fully described in a paper presented to the Royal Society by Mr. Buchanan early in 1875, and published in abstract in the ' Proceedings' for that year. Its description is briefly as follows:—

The stem, which carries a millimetre-scale 10 centimetres long, has an outside diameter of about 3 millimetres, the external volume of the divided portion being 0·8607 cubic centimetre ; the mean volume of the body is 160·15 cubic centimetres, and the weight of the glass instrument is 160·0405 grammes. With this volume and weight it floats in distilled water of 16° C., at about the lowest division (100) of the scale. In order to make it serviceable for heavier waters, a small brass table is made to rest on the top of the stem, of such a weight that it depresses the instrument in distilled water of 16° C. to about the topmost division (0) of the scale. By means of a series of six weights, multiples by 1, 2, 3, 4, 5, and 6 of the weight of the table, specific gravities between 1·00000 and 1·03400 can be observed. It is not necessary that these weights should be accurate multiples of the weight of the table; it is sufficient if they approach it within a centigramme, and their actual weight be known with accuracy. The weights of the table and weights in actual use are :—

Weight of table			0·8360	gramme.
,,	of weight No.	I.	0·8560	,,
,,	,,	II.	1·6010	,,
,,	,,	III.	2·4225	grammes.
,,	,,	IV.	3·1245	,,
,,	,,	V.	4·0710	,,
,,	,,	VI.	4·8245	,,

For ocean-waters the hydrometer is always used with the table and either No. IV. or No. V. weight.

When the mechanical part of the construction of the instrument was finished, with the exception of the closing of the top of the stem (which instead was widened into a funnel-shape large enough to receive

A A 2

the ordinary decigramme weights), the calibration
of the stem was effected by loading the stem with
successive weights, and observing the consequent
depressions in distilled water of known temperature.
This done, the top was sealed up and the instrument
carefully weighed. The expansion of the body with
temperature was determined in a similar manner by
reading the instrument in distilled water of various
temperatures. The coefficient of expansion of the
glass was then found to be 0·000029 per degree
Centigrade.

For using this instrument at sea about 900 cubic
centimetres of sea-water are taken, and the containing cylinder placed on a swinging table in a position
as near the centre of the ship as possible. The observation with the hydrometer, loaded with the
necessary table and weight, is then effected in the
ordinary way, the accuracy of the readings being
but little affected by rolling ; pitching, however,
is found to have a distinctly disturbing effect; and
when it is in any way violent, it is advisable to store
the specimen of water till the weather improves.

The temperature of the water at the time of
observation is determined by one of Geissler's
'normal' or standard thermometers, graduated into
tenths of a degree Centigrade; and it is essential
for the accuracy of the results that the water, during
the observation of the hydrometer, should be sensibly
at the same temperature as the atmosphere, otherwise
the changing temperature of the water makes the
readings of both the hydrometer and the thermometer
uncertain. At low temperatures (below 10° or 12° C.)
a tenth of a degree makes no sensible difference in

the resulting specific gravity ; but at the high temperatures always found at the surface of tropical seas,
rising sometimes to 30° C., the same difference of
temperature may make a difference of 3 to 4 in the
resulting specific gravity.

Having obtained the specific gravity of the water
in question at a temperature which depends upon
that of the air at the time, it is necessary, in order
that the results may be comparable, to reduce them
to their values at one common temperature. For
this purpose a knowledge of the law of expansion
of sea-water with temperature is necessary. This
had been determined with sufficient accuracy for
low temperatures by Despretz and others ; but as
the temperatures at which specific-gravity observations are usually made are comparatively high, their
results were of but little use, directed as they were
chiefly to the determination of the freezing and
maximum-density points. When the late Captain
Maury was developing his theory of oceanic circulation, owing to difference of density of the water
in its different parts, he found the want of information on this important subject. At his request
the late Professor Hubbard, of the National Observatory, U.S., instituted a series of experiments, from
which he was enabled to lay down a curve of the
volumes of sea-water at all temperatures from considerably below the freezing-point to much above
what obtains even in the hottest seas. The results
are published in Maury's *Sailing Directions*, 1858,
vol. i. p. 237, and have evidently been carried out
with great care. The composition of different oceanic
waters varies, even in extreme cases, within such

close limits, that the law of thermal expansion is
sensibly the same for all of them; of this Hubbard's
experiments afford satisfactory proof. In the Table
which gives the results of all his experiments he
takes the volume of water at 60° F. as his unit.

In the following Table the volumes for every
Centigrade degree from − 1° C. to + 30° C. are
given :—

Temp. °C.	Volume.	Temp. °C.	Volume.	Temp. °C.	Volume.	Temp. °C.	Volume.
−1	0·99792	+7	0·99853	+15	0·99987	+23	1·00194
0	795	8	866	16	1·00010	24	224
+1	799	9	878	17	034	25	256
2	804	10	893	18	059	26	288
3	812	11	910	19	086	27	320
4	820	12	927	20	111	28	352
5	830	13	947	21	137	29	385
6	840	14	967	22	164	30	420

The results of Mr. Buchanan's observations are
given for each section in Tables forming Appendices
to the several chapters. In these Tables the specific
gravity is given at the temperature at which the determination was made; at the temperature reduced
to 15°·56 C.; and at the temperature which it had
in the position in the ocean from which it was
taken ;—the specific gravity of distilled water at
4° C. being retained as the unit.

With a single exception, off the coast of Brazil, the
densest water which we have met with in the ocean
was found on the section from Teneriffe to St.
Thomas's in the heart of the north-east trade-wind
territory, where, from the strength and dryness of the
wind, the amount of evaporation must be very large.

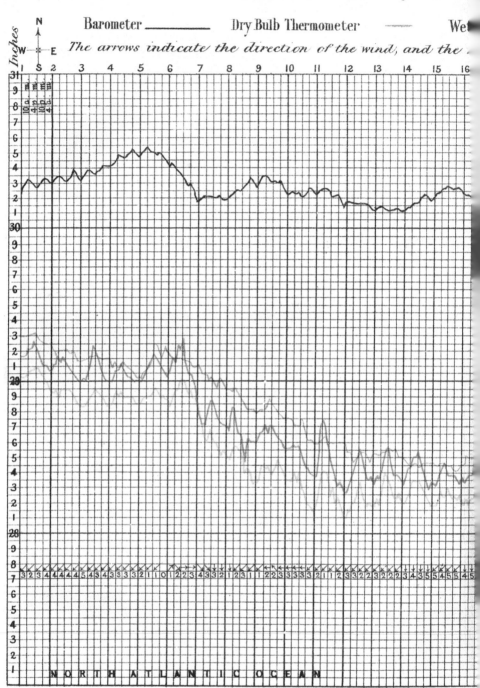

PLATE XLII. *Meteorological Obser*

Barometer ——————— Dry Bulb Thermometer — — Wet

The arrows indicate the direction of the wind, and the .

NORTH ATLANTIC OCEAN

b Thermometer　　　　　　　Temperature of Sea Surface

bers beneath its force according to Beaufort's scale

7　18　19　20　21　22　23　24　25　26　27　28　29　30　31　40 °C

40

5

30

5

20

5

5

10

4 4 4 4 5 7 6 5 5 5 5 6 6 7 1 0 0 1 1 3 3 2 3 4 5 5 4 3 1 2

VIGO HR　　　　PORTSMOUTH

Round about the Canary Islands the mean specific gravity was found to be 1·02730; to the westward it rises steadily until in longitude 28° W. it has reached 1·02762. Between longitude 28° W. and 54° W. the mean specific gravity is 1·02773 the maximum being 1·02781. On approaching the West Indies it rapidly falls off to an average of 1·02719 in the neighbourhood of St. Thomas's; and if we take into account all the observations made on the western side of the Atlantic, from St. Thomas's northward to the edge of the cold water which separates the Gulf-stream from the coast of America, we obtain the same average, 1·02719. Between Bermudas and the Azores an almost perfectly uniform specific gravity was observed, the mean being 1·02713, and the extremes 1·02694 and 1·02727. As Madeira is approached the specific gravity rises until it reaches 1·02746 close to the island itself. The mean specific gravity on the eastern side of the North Atlantic, between the latitude of St. Thomas's and that of the Azores, is 1·02727, or slightly higher than that of the water on the western side.

After leaving the Cape-Verde Islands, the ship's course lay almost parallel to the African coast, and at an average distance of about 200 miles from it. Proceeding thus in a south-easterly direction, the specific gravity fell rapidly from 1·02692 off St. Iago on the 10th August, to 1·02632 on the 12th, after which it retained the low mean specific gravity of 1·02627 until the 21st August, when the course was changed to a westerly one along the equator. The specific gravity of the water on this day was the lowest hitherto registered for a surface-water; it was

1·02601, in lat. 3° 8' N., and on the boundary line between the equatorial and Guinea currents. The same low specific gravity was observed in following the equatorial current as far as St. Paul's rocks, after which it quickly rose as the Brazilian coast was approached; and the maximum of 1·02786 was obtained on the 26th September, when off the entrance to Bahia, in latitude 13° 4' S.

The observations in the South Atlantic were limited to a line down the western side as far as the Abrolhos Bank, and thence across to the Cape of Good Hope. In the region of the south-east trade-wind, therefore, we have only a few observations close to the coast; and as we have seen in the North Atlantic, on the voyage from Teneriffe to St. Thomas's, the specific gravity is higher in mid-ocean than either on the east or the west side, so in the South Atlantic it is possible that the same may hold good. From the Abrolhos Bank to Tristan d'Acunha the specific gravity sinks steadily from 1 02785 to 1·02606, and from Tristan to the Cape of Good Hope, along a course lying between the 35th and the 37th parallels of south latitude, the mean specific gravity was 1·02624. Between the same parallels of north lati-tude the mean specific gravity was 1·02713.

It must be remembered that the results obtained can only be held good for the season of the year in which they were observed, and that the observations in different latitudes were made in different seasons, and, further, that all the observations north of the line as far as 20° N. were obtained on the eastern side, and those to the southward of it as far as 30° S. were obtained on the western side of the ocean; so that it would be unwise to attempt to draw any general conclusions from such imperfect data. Considering, however, our four parallel sections, we have at least this positive result—that in the month of June and mean latitude 36° N. the surface-water in mid-ocean has a mean specific gravity of 1·02712, that in the months of February and March and mean latitude 22° N. the mean surface specific gravity is 1·02773, that in the month of August and mean latitude 2° N. it is 1·02624, and that in the month of October in mean latitude 36′ S. it is 1·02621.

On the way to and from Halifax in the month of May some observations were obtained in the cold water with which the north-eastern coast of America is surrounded, the mean specific gravity being 1·02463. On the 1st May in the Gulf-stream the specific gravity of the water was 1·02675, and its temperature 23°·9 C.; and the next day it was 1·02538, and the temperature 13°·3 C. If the results be reduced to their values at the respective temperatures of the different waters, we have for the specific gravity of the Gulf-stream water 1·02445, and of Labrador-current water 1·02584 ; so that the fall of temperature very much more than counterbalances the want of salt in the water. In the same way we find the mean specific gravity of the water referred to the temperature which it has in the ocean to be—in latitude 36° N. and month of June 1·02548, in 22° N. and month of February and March 1·02592, in 2° N. and month of August 1·02335, and in 36° S. and month of October 1·02659.

From the determination of the specific gravity of intermediate and bottom-water, Mr. Buchanan concludes that, as a general rule, both in the Atlantic and the Pacific Oceans, between the parallels of 40° N. and 40° S., the specific gravity reduced to 15°·56 C., is greatest at or near the surface, and decreases more or less regularly until a minimum is reached, generally 400 fathoms from the surface, whence there is a slow rise, the bottom-water being slightly heavier.

From Mr. Buchanan's report, and from the specific gravity Tables of the year 1873, we come then to the broad conclusion that the density of the upper-layers of the North Atlantic is considerably higher than that of any other part of the ocean, and the specific gravity tables for the spring months of 1876 give the same result. I need scarcely say that this is exactly what must have been anticipated, if my view be correct, that the movement of deep-water in the Atlantic is mainly due to excess of evaporation over precipitation in its northern portion.

An element of great uncertainty is undoubtedly introduced into the determination of the specific gravity of surface by the weather. These determinations were made as usual daily for the section between Stations 323 and 335 from Monte Video to Tristan d'Acunha in the beginning of March, 1876, and the mean of these, the temperature reduced to 15°·56 C., was 1·02620. Of the eighteen days occupied in running the section, nine were dry and fine, and on nine rain fell either continuously or in showers. The mean for the nine dry days was 1·02639, and for the nine wet days 1·02591. The maximum surface specific gravity for the section (1·02680) was at Station 323, at the point where probably the Brazil current has most effect on the surface, and the minimum (1·02494) was at Station 326 after a heavy fall of rain. The mean specific gravity of the surface-water at the temperature at which it was procured was 1·02502.

The specific gravity of the bottom-water was determined at ten stations on the section. Reduced to a temperature of 15°·56 C., the mean was 1·02601; the maximum, 1·02650, was at Station 323 at a depth of 1,900 fathoms; and the minimum. 1·02580, was at Station 326 at 2,775 fathoms. The mean specific gravity of the bottom-water at the depth at which it was procured was 1·02811, showing a difference between the two means of 0·00210, due to difference of temperature alone.

It seems from these observations that the differences of surface specific gravity due to differences of salinity along the section are very small, and that, with the exception possibly of Station 323, which is abnormal in many respects, they depend mainly on the rainfall.

The difference between the mean surface specific gravity, the temperature reduced to 15°·56 C., and the mean bottom specific gravity under the same conditions is also very slight; the actual specific gravity at every point is practically determined by the temperature; and consequently the bands of equal density are, like the bands of equal temperature, virtually continuous with those of the Southern Sea.

The Amount of Carbonic Acid contained in Sea-water.—I give in Appendix C. to this chapter a table of carbonic acid determinations from Mr. Buchanan's 'Laboratory Work;' and the substance of the few following remarks on the subject is taken from his preliminary report ('Proceedings of the Royal Society,' vol. xxiv., p. 602 *et seq.*)

The carbonic acid when boiled out of the water was received by baryta-water of known strength; its consequent loss of alkalinity was measured by hydrochloric acid of corresponding strength. Having observed that the presence of sulphates in sea-water is one of the potent agents in the retention of the carbonic acid ('Proceedings of the Royal Society,' vol. xxii., p. 483 *et seq.*), Mr. Buchanan always added 10 cubic centimetres of a saturated solution of chloride of barium to the water before commencing the operation. This facilitates greatly the liberation of the carbonic acid, and also causes the water to boil tranquilly, even to dryness, without showing any tendency towards bumping. The quantity of water used has been almost invariably 225 cubic centimetres, and the property possessed by sea-water of retaining its carbonic acid with great vigour makes it possible to perform the determination of it even a couple of days after its collection.

As in the great majority of cases, where the carbonic acid has been determined, the oxygen and nitrogen have also been collected, and have been preserved until our return home, where they will shortly be analyzed: it would be useless to attempt to discuss the results of the carbonic-acid determinations at present, and before these analyses have been made, especially as there is likely to be some relation between the amounts of oxygen and of carbonic acid. Independently, however, of the relations which may subsist between the two bodies, it may be gathered from the inspection of the table (Appendix C.) that, taking surface-waters alone, the amount of carbonic acid present is many times greater than would be contained in the same volume of distilled water under the same circumstances. Sometimes it is more than thirty times as much.

The amount of carbonic acid contained by surface-waters of the same temperatures increases with the density, and consequently is greater in the surface-water of the Atlantic than in that of the Pacific, the two oceans being very markedly distinguished from one another by the different densities of their surface-waters. Thus we have a mean of 0·0466 gramme CO_2 per litre in Atlantic surface-water of temperature between 20° and 25° C. and mean density 1·02727; whilst in the Pacific the mean is 0·0268 gramme in water of 1· 2594 mean density; and the mean amount of carbonic acid in Atlantic water of temperature above 25° C. and mean density 1·02659 is 0·0409, whilst in the Pacific the corresponding water is of mean density 1·02593, and contains 0·0332 gramme CO_2 per litre. As a rule, other things being equal, the amount of carbonic acid diminishes as the temperature increases; thus the mean amount of carbonic acid in waters whose temperature was between 15° and 20° was found to be 0·0446 gramme per litre, the mean density being 1·02642, whilst we have seen that in the Atlantic the surface-water of temperature above 25° C. and of mean density 1·02659 contains 0·0409 gramme per litre. Also there is usually more carbonic acid in waters taken from the bottom and intermediate depths than in surface-water; but if regard be had to the temperature of the water, it will be seen that there is but little difference in the amount in waters of the same temperature from whatever depth they may have been derived. This seems to indicate that the animal life at the bottom and at great depths cannot be very abundant, otherwise there could hardly fail to be a decided excess of carbonic acid in the deep water, owing to constant production and want of the means of elimination of the gas. On this subject, however, it would be premature to speculate before the determination of the oxygen, from which we may hope for much information.

At a meeting of the Royal Society of Edinburgh, on the 4th of June of the present year, Mr. Buchanan communicated the results of an examination of the gases dissolved in sea-water at different depths, especially with reference to the amount of oxygen contained.

He finds that at the surface the amount of oxygen varies between 33 and 35 per cent., the higher number having been observed in a water collected almost on the Antarctic circle; the smallest percentages have been observed in the trade-wind districts. In bottom-water, the absolute amount is greatest in Antarctic regions, diminishing generally towards the north. The oxygen percentage is greatest over 'diatomaceous oozes,' and least over 'red-clays' containing peroxide of manganese; over 'blue-clays' it is greater than over 'globigerina oozes.' In intermediate waters the remarkable fact was observed that the oxygen diminishes down to a depth of 300 fathoms, at which point it attains a minimum, after which the amount increases.

The following figures show the nature of this phenomenon :—

Depth (fathoms).	0	25	50	100	200	300	400	800	Between 800 and the bottom.[1]
Oxygen $O + N = 100$	33·7	33·4	32·2	30·2	33·4	11·4	15·5	22·6	23·5

[1] *Nature*, July 26th, 1877.

Mr. Buchanan drew the conclusion in explanation of the small amount of oxygen at depths of 300 fathoms and upwards, 'that animal life must be particularly abundant and active at this depth, or at least more abundant than at greater depths.' In other words, that a permanent condition, probably of all conditions the most unfavourable to animal life, is produced and maintained by its excess.

This is entirely contrary to experience. I think, however, that the observation, which is in itself of the highest interest, goes far to support the opposite opinion, at which I had previously arrived from other considerations, that in deep water a wide intermediate zone between the surface and the layer immediately above the bottom is nearly destitute of animal life, at all events in its higher manifestations.

If the view which I have adopted of the cause and course of the circulation of the water in the Atlantic and Pacific Oceans be correct, it seems to afford a ready explanation of the peculiar distribution of oxygen. Free oxygen is doubtless in all cases derived by the water of the sea from the atmosphere, and it is consequently absorbed through the surface where the water is constantly agitated in contact with the air, and the surface-water contains most.

In the Antarctic regions, the surface-water sinks rapidly to the bottom, and moves northwards as the cold southern indraught. The bottom-water has thus, next to the surface-water, had the latest opportunity of becoming impregnated with air, and a considerable portion of that air it retains. If the deep circulation in the Atlantic and the Pacific be chiefly maintained, as I have been led to believe, by evaporation of the surface-water and a slow indraught of Antarctic water beneath to supply its place, a central belt, or at all events a belt at too great a depth to be affected by surface influences, must be the *oldest* water in the vertical section, and must consequently have been longest subjected to the removal of oxygen by the scanty fauna which may still subsist, and more especially by the oxidation of the products of the decomposition of surface organisms as they sink through it towards the bottom.

A great deal has yet to be done before we can be in a position to generalize with safety on the many chemical questions of great interest which have been raised during the progress of the Expedition. I hope, however, that the next two years may see the water- and gas-analyses, and the analyses of the matters of mineral and organic origin which form the deep-sea deposits, well advanced; and that the complete data in this department may appear in the form of appendices to an early volume of the official report.

APPENDIX A.

The General Results of the Chemical and Microscopical Examination of a Series of Twenty Samples of the Bottom, from the Observing Stations on the Section between Teneriffe and Sombrero.

(The samples were analysed by Mr. James S. Brazier, Regius Professor of Chemistry in the University of Aberdeen; the microscopic work was done by Mr. John Murray, and the results are taken from his notes.)

No. 1.—Station 1. February 15th, 1873. Lat. 27° 24′ N. Long. 16° 55′ W. Depth, 1,890 fathoms. Bottom temperature 2°·0 C. Chemical composition :—

Loss on ignition after drying at 230° F.		7·91
	Alumina	5·26
	Ferric oxide	3·95
Portion soluble in	Calcium phosphate	Large traces:
hydrochloric acid	Calcium sulphate	0·44
= 73·07.	Calcium carbonate	.50·00
	Magnesium carbonate	1·32
	Silica	12·10
Portion insoluble	Alumina	} 3·47
in hydrochloric	Ferric oxide	}
acid = 19·02.	Lime	1·26
	Magnesia	0·52
	Silica	13·77
		100·00

A 'globigerina-ooze,' containing many coccoliths and rhabdoliths, many pelagic foraminifera of the genera *Globigerina, Pulvinulina, Orbulina, Pullenia,* &c.

VOL. II.   B B

Amorphous clayey and calcareous matter; and small particles of felspar, mica, quartz, hornblende, and magnetite.

No. 2.—Station 2. February 17th. Lat. 25° 52′ N.; Long. 19° 14′ W. Depth, 1,945 fathoms. Bottom temperature, 2°·0 C. Chemical composition :—

Loss on ignition after drying at 230° F.		5·02
Portion soluble in hydrochloric acid = 82·90.	Alumina	3·23
	Ferric oxide	4·18
	Calcium phosphate	Trace.
	Calcium sulphate.	0·69
	Calcium carbonate	64·55
	Magnesium carbonate	1·17
	Silica	9·08
Portion insoluble in hydrochloric acid = 12·08.	Alumina	1·79
	Ferric oxide	0·60
	Lime	0·33
	Magnesia	0·28
	Silica	9·08
		100·00

A 'globigerina-ooze' of a grey colour, containing many pelagic foraminifera of the genera *Globigerina*, *Pulvinulina*, *Orbulina*, *Pullenia*, and *Sphæroidina*. A few *Biloculinæ* and arenaceous foraminifera. A few shells of pteropods, otolites of fishes, and spines of echini. A few spicules of sponges and radiolarians.

Amorphous clayey matter, and many small particles of quartz, mica, magnetite, felspar, and augite. The larger mineral particles were rounded as if wind-blown.

No. 3.—Station 5. February 21st. Lat. 24° 20′ N.; Long. 24° 28′ W. Depth, 2,740 fathoms. Bottom temperature, 2°·0 C. Chemical composition :—

Loss on ignition after drying at 230° F.		8·20
Portion soluble in hydrochloric acid = 77·30.	Alumina	4·70
	Ferric oxide	3·50
	Calcium phosphate	Traces.
	Calcium sulphate	0·70
	Calcium carbonate	56·39
	Magnesium carbonate	0·98
	Silica	11·03

Portion insoluble in hydrochloric acid = 14·50.	Alumina	1·80
	Ferric oxide	0·80
	Lime	0·50
	Magnesia	0·40
	Silica	11·00

100 00

A 'red clay,' containing many pelagic foraminifera of the genera *Globigerina, Orbulina, Sphæroidina, Pullenia*, and *Pulvinulina*; a few *Biloculinæ* and arenaceous foraminifera; a few radiolaria, and one or two pteropod shells.

Much amorphous clayey matter deeply dyed with oxide of iron. Many small mineral particles—mica, magnetite, felspar, quartz, and hornblende. These mineral particles appeared wind-blown, and had probably been carried to this area by the Harmattan and trade-winds.

No. 4.—Station 7. February 24th. Lat. 23° 23′ N.; Long. 31 31′ W. Depth, 2,750 fathoms. Bottom temperature 2°·0 C. Chemical composition :—

Loss on ignition after drying at 230° F.		7·45
Portion soluble in hydrochloric acid = 52·98.	Alumina	6·40
	Ferric oxide	15·42
	Calcium phosphate	Trace.
	Calcium sulphate	1·60
	Calcium carbonate	4·11
	Magnesium carbonate	1·20
	Silica	24·25
Portion insoluble in hydrochloric acid = 39·57.	Alumina	6·00
	Ferric oxide	2·54
	Lime	1·06
	Magnesia	0·64
	Silica	29·33

100·00

A 'red clay,' containing much amorphous clayey matter, and many small mineral particles—quartz, mica, hornblende, felspar, magnetic iron. A few broken pieces of pelagic foraminifera.

No. 5.—Station 8. February 25th. Lat. 23° 12′ N.; Long. 32° 56′ W. Depth, 2,800 fathoms. Bottom temperature, 2°·0 C. Chemical composition :—

Loss on ignition after drying at 230° F. . . . . 8·95

	Alumina	8·95
	Ferric oxide	9·70
Portion soluble in	Calcium phosphate	Large trace.
hydrochloric acid	Calcium sulphate	2·24
= 63·01.	Calcium carbonate	16·42
	Magnesium carbonate	2·70
	Silica	23·00
	Alumina	4·20
Portion insoluble	Ferric oxide	2·10
in hydrochloric	Lime	0·89
acid = 28·04.	Magnesia	0·60
	Silica	20·25

100·00

A 'red clay,' containing much amorphous clayey matter, and many fine mineral particles—mica, quartz, felspar, magnetite, and augite. A few pelagic foraminifera entire and broken; a few arenaceous foraminifera.

No. 6.—Station 9. February 26th. Lat. 23° 23′ N.; Long. 35° 10′ W. Depth, 3,150 fathoms. Bottom temperature, 1°·9 C· Chemical composition :—

Loss on ignition after drying at 230° F. . . . . 10·40

	Alumina	8·30
	Ferric oxide	9·75
Portion soluble in	Calcium phosphate	Good traces.
hydrochloric acid	Calcium sulphate	0·87
= 43·74.	Calcium carbonate	3·11
	Magnesium carbonate	1·90
	Silica	19·81
	Alumina	9·10
Portion insoluble	Ferric oxide	2·04
in hydrochloric	Lime	0·47
acid = 45·86.	Magnesia	0·95
	Silica ·	33·30

100·00

A 'red clay' containing much amorphous clayey matter, many particles of mica, magnetite, quartz, and hornblende; some of the larger particles were rounded.

A very few broken portions of pelagic foraminifera occurred, and a few arenaceous forms.

No. 7.—Station 10. February 28th. Lat. 23° 10′ N. ; Long. 38° 42′ W. Depth, 2,720 fathoms. Bottom temperature, 1°·9 C. Chemical composition :—

Loss on ignition after drying at 230° F.		7·61
	Alumina	9·73
	Ferric oxide	9·30
Portion soluble in	Calcium phosphate	—
hydrochloric acid	Calcium sulphate	0·61
= 58·98.	Calcium carbonate	13·30
	Magnesium carbonate	1·31
	Silica	24·73
	Alumina	5·50
Portion insoluble	Ferric oxide	2·96
in hydrochloric	Lime	0·23
acid = 33·41.	Magnesia	0·19
	Silica	24·53
		100·00

A ' red clay ' containing much amorphous clayey matter and many fine mineral particles—felspar, mica, quartz, and magnetite.

A few entire and many broken pelagic and arenaceous foraminifera.

No. 8.—Station 11. March 1st. Lat. 22° 45′ N. ; Long. 40° 37′ W. Depth, 2,575 fathoms. Bottom temperature, 2°·0 C. Chemical composition :—

Loss on ignition after drying at 230° F.		9·13
	Alumina	5·61
	Ferric oxide	4·65
Portion soluble in	Calcium phosphate	—
hydrochloric acid	Calcium sulphate	1·02
= 76·59.	Calcium carbonate	51·16
	Magnesium carbonate	1·93
	Silica	12·22
Portion insoluble in hydrochloric acid = 14·28.	Insoluble residue, principally alumina and ferric oxide, with silica	14·28
		100·00

A ' red clay ' containing much deep red amorphous clayey matter, with many particles of felspar, magnetite, augite, mica, quartz, &c.

A good many pelagic foraminifera and their fragments. Coccoliths and rhabdoliths.

No. 9.—Station 12. March 3rd. Lat. 21° 57′ N.; Long. 43° 29′ W. Depth, 2,025 fathoms. Bottom temperature, 1°·9 C. Chemical composition :—

Loss on ignition after drying at 230° F. . . . . 8·80
Alumina . . . . 19·24
Ferric oxide . . . . 13·74
Calcium phosphate . . . . Fair traces.
Calcium sulphate . . . . 1·37
Calcium carbonate . . . . 43·93
Magnesium carbonate . . . . 1·94
General residue, consisting of soluble silica with the insoluble
silicates . . . . 10·98
_____
100·00

A 'globigerina-ooze' containing many pelagic foraminifera of the genera *Globigerina, Orbulina, Pulvinulina, Sphæroidina,* and *Pullenia.* Many coccoliths and rhabdoliths.

Much amorphous clayey matter, with iron and manganese peroxides.

No. 10.—Station 13. March 4th. Lat. 21° 38′ N.; Long. 44° 39′ W. Depth, 1,900 fathoms. Bottom temperature, 1°·9 C. Chemical composition :—

Loss on ignition after drying at 230° F. . . . . 6·63
Alumina . . . . } 5·86
Ferric oxide . . . . }
Calcium phosphate . . . . Small traces.
Calcium sulphate . . . . 0·51
Calcium carbonate . . . . 74·50
Magnesium carbonate . . . . 1·27
General residue, consisting of soluble silica with the insoluble
silicates . . . . 11·23
_____
100·00

A 'globigerina-ooze' containing many pelagic foraminifera of the genera *Globigerina, Hartigerina, Pulvinulina, Sphæroidina,* and *Orbulina.* Many coccoliths and rhabdoliths. A few pteropod shells and valves of Ostracoda and otolites of fishes.

Amorphous clayey matter and small mineral particles—mica, quartz, olivine, felspar, and pumice. Some of the particles of quartz were rounded as if wind-blown.

No. 11.—Station 14. March 5th. Lat 21° 1′ N.; Long. 46° 29′ W. Depth, 1,950 fathoms. Bottom temperature, 1°·8 C. Chemical composition :—

Loss on ignition after drying at 230° F.		4·58
Portion soluble in hydrochloric acid = 90·82.	Alumina . . . . } Ferric oxide . . . . }	3·33
	Calcium phosphate	1·12
	Calcium sulphate	1·20
	Calcium carbonate	79·17
	Magnesium carbonate	1·40
	Silica	4·60
Portion insoluble in hydrochloric acid = 4·60.	Insoluble residue, principally alumina and ferric oxide, with silica	4·60

	100·00

A reddish 'globigerina-ooze' containing many pelagic foraminifera of the usual genera, and many coccoliths and rhabdoliths.

Amorphous clayey matter with oxide of iron. Many small particles of sanidine, augite, hornblende, and magnetite.

No. 12.—Station 15. March 6th. Lat. 20° 49′ N.; Long. 48° 45′ W. Depth, 2,325 fathoms. Bottom temperature, 1°·7 C. Chemical composition :—

Loss on ignition after drying at 230° F.		4·17
Portion soluble in hydrochloric acid = 87·50.	Alumina . . . . } Ferric oxide . . . . }	6·25
	Calcium phosphate . . . . Large traces.	
	Calcium sulphate	1·91
	Calcium carbonate	67·60
	Magnesium carbonate	2·58
	Silica	9·16
Portion insoluble in hydrochloric acid = 8·33.	Insoluble residue, principally alumina and ferric oxide, with silica	8·33

	100·00

A 'globigerina-ooze,' containing many pelagic foraminifera of the genera *Globigerina, Orbulina, Pulvinulina,* and *Sphæroidina;* many coccoliths and rhabdoliths.

Amorphous clayey matter with oxide of iron. Small particles of sanidine, augite, pumice, magnetite, &c. A few grains of manganese peroxide.

No. 13.—Station 16. Lat. 20° 39′ N.; Long. 50° 33′ W. Depth, 2,435 fathoms. Bottom temperature, 1°·7 C. Chemical composition :—

Loss on ignition after drying at 230° F.		9·60
	Alumina	4·00
	Ferric oxide	7·10
Portion soluble in	Calcium phosphate	Small traces.
hydrochloric acid	Calcium sulphate	2·32
= 78·40.	Calcium carbonate	52·22
	Magnesium carbonate	0·76
	Silica	12·00
	Alumina	} 2·96
Portion insoluble	Ferric oxide	}
in hydrochloric	Lime	0·64
acid = 12·00.	Magnesia	0·40
	Silica	8·00
		100·00

A 'red clay,' containing amorphous clayey matter with oxide of iron, and many small particles of magnetite, felspar, pumice, and hornblende. A few grains of manganese peroxide.

Many pelagic foraminifera of the genera *Globigerina, Orbulina, Sphæroidina,* and *Pulvinulina.* Coccoliths and rhabdoliths. The dredge brought up five small round manganese concretions about the size of marbles, and three sharks' teeth of the genus *Lamna* with a slight coating of manganese peroxide.

No. 14.—Station 17. Lat. 20° 7′ N.; Long. 52° 32′ W. Depth, 2,385 fathoms. Bottom temperature, 1°·9 C. Chemical composition :—

Loss on ignition after drying at° 230 F. . . . .   6·84

Portion soluble in hydrochloric acid = 83·44.
- Alumina . . . .   2·69
- Ferric oxide . . . .   9·05
- Calcium phosphate . . . .   1·74
- Calcium sulphate. . . . .   0·81
- Calcium carbonate . . . .   58·40
- Magnesium carbonate . . . .   0·68
- Silica . . . .   10·07

Portion insoluble in hydrochloric acid = 9·72.
- Insoluble residue, principally alumina and ferric oxide, with silica . . . . . .   9·72

100·00

A 'red-clay' containing amorphous clayey matter with oxide of iron, and many small particles of sanidine, augite, magnetite, and quartz. A few grains of manganese peroxide. Many pelagic foraminifera of the genera *Globigerina*, *Pulvinulina*, *Sphæroidina*, &c. ; coccoliths and rhabdoliths.

No. 15. Station 18. March 10th. Lat. 19° 41′ N ; Long. 55° 13′ W. Depth, 2,675 fathoms. Bottom temperature, 1°·6 C. Chemical composition :—

Loss on ignition after drying at 230° F. . . . .   7·75

Portion soluble in hydrochloric acid = 60·00.
- Alumina . . . .   8·25
- Ferric oxide . . . .   11·37
- Calcium phosphate . . . .   0·42
- Calcium sulphate . . . .   0·52
- Calcium carbonate . . . .   15·78
- Magnesium carbonate . . . .   1·41
- Silica . . . .   22·25

Portion insoluble in hydrochloric acid = 32·25.
- Alumina . . . .   7·00
- Ferric oxide . . . .   2·50
- Lime . . . .   0·57
- Magnesia . . . .   0·38
- Silica . . . .   21·80

100·00

A 'red-clay' containing amorphous clayey matter, and small particles of augite, felspar, hornblende, and magnetite. A few grains of manganese peroxide. A few broken tests of pelagic foraminifera, coccoliths and rhabdoliths.

No. 16.—Station 19. March 11th. Lat. 19° 15′ N.; Long. 57° 47′ W. Depth, 3,000 fathoms. Bottom temperature, 1°·3 C. Chemical composition :—

Loss on ignition after drying at 230° F.		7·44
	Alumina	12·91
	Ferric oxide	10·33
Portion soluble in	Calcium phosphate	Traces.
hydrochloric acid	Calcium sulphate	0·96
= 56·47.	Calcium carbonate	1·49
	Magnesium carbonate	3·10
	Silica	27·68
	Alumina	7·81
Portion insoluble	Ferric oxide	1·57
in hydrochloric	Lime	1·03
acid = 36·09.	Magnesia	0·52
	Silica	25·16
		100·00

A ' red-clay ' containing amorphous clayey matter with oxide of iron. Small crystals of sanidine, mica, augite.

A few siliceous spicules. Only a single fragment of *Globigerina* shell was observed.

No. 17.—Station 20. March 12. Lat. 18° 56′ N.; Long. 59° 35′ W. Depth, 2,975 fathoms. Bottom temperature, 1°·6 C. Chemical composition :—

Loss on ignition after washing and drying at 230° F.		7·45
	Alumina	12·28
	Ferric oxide	11·44
Portion soluble in	Calcium phosphate	Small trace.
hydrochloric acid	Calcium sulphate	1·47
= 56·83.	Calcium carbonate	3·50
	Magnesium carbonate	2·14
	Silica	26·00
	Alumina	7·28
Portion insoluble	Ferric oxide	2·36
in hydrochloric	Lime	1·18
acid = 35·72.	Magnesia	0·50
	Silica	24·40
		100·00

A 'red clay' containing amorphous clayey matter with oxide of iron; small particles of hornblende, augite, magnetite, sanidine, and quartz, and a few grains of peroxide of manganese.

A few siliceous spicules. Only two fragments of *Globigerina* shell occurred in the portion of the sample examined.

No. 18.—Station 21. March 13. Lat. 18° 54′ N.; Long. 61° 28′ W. Depth, 3,025 fathoms. Bottom temperature, 1°·3 C. Chemical composition:—

Loss on ignition after drying at 230° F.		5·92
Portion soluble in hydrochloric acid = 50·42.	Alumina	7·04
	Ferric oxide	12·25
	Calcium phosphate	Small traces.
	Calcium sulphate	0·51
	Calcium carbonate	2·44
	Magnesium carbonate	3.48
	Silica	24·70
Portion insoluble in hydrochloric acid = 43·66.	Alumina	5·51
	Ferric oxide	6·73
	Lime	0·81
	Magnesia	0·41
	Silica	30·20
		100·00

A 'red clay' containing much amorphous clayey matter with iron peroxide; many fragments of sanidine, augite, olivine, hornblende, and magnetite; many of the mineral particles much larger than those at Station 20.
A few fragments of the tests of *Globigerina*.

No. 19.—Station 22. March 14.—Lat. 18° 40′ N.; Long. 62° 56′ W. Depth, 1,420 fathoms. Bottom temperature, 3°·0 C. Chemical composition:—

Loss on ignition after drying at 230° F.		3·80
Portion soluble in hydrochloric acid = 92·75.	Alumina	4·42
	Ferric oxide	
	Calcium phosphate	2·41
	Calcium sulphate	0·41
	Calcium carbonate	80·69
	Magnesium carbonate	0·68
	Silica	4·14
Portion insoluble in hydrochloric acid = 3·45.	Insoluble residue, principally alumina and ferric oxide, with silica	3·45
		100·00

A 'globigerina-ooze' containing many pelagic foraminifera of the genera *Globigerina, Orbulina, Pulvinulina, Pullenia,* and *Sphæroidina.*

Many shells of pteropods and heteropods; a few coccoliths and rhabdoliths; otolites of fishes, and spines of echini; a few siliceous spicules.

Amorphous mineral matter and particles of quartz, felspar, hornblende, and magnetite.

No. 20.—Station 23.   March 15.   Off Sombrero Island. Depth 450 fathoms.   Chemical composition:—

Loss on ignition after drying at 230° F.	. . . .	4·00
	Alumina	1·80
	Ferric oxide	3·00
Portion soluble in	Calcium phosphate	Good traces.
hydrochloric acid	Calcium sulphate	1·00
= 93·95.	Calcium carbonate	84·27
	Magnesium carbonate	1·28
	Silica	2·60
Portion insoluble in hydrochloric acid = 2·05.	Insoluble residue, principally alumina and ferric oxide, with silica	2·05
		100·00

A 'pteropod-ooze' containing very many shells of pteropods and heteropods and their broken fragments; many pelagic foraminifera of the genera *Globigerina, Pulvinulina, Orbulina, Pullenia,* and *Sphæroidina;* large *Biloculinæ* and calcareous *Rotaliæ* and *Cristellariæ;* a few coccoliths.

Amorphous clayey and calcareous matter, with sandy particles, quartz, felspar, mica, magnetite, and sanidine.

Notes on the foregoing analyses by Professor Brazier.

The loss on ignition consists for the most part of water, probably water of hydration; but there is in all cases evidence of the existence of organic matter. The majority of the specimens, when treated with hydrochloric acid, evolved the peculiar tarry odour so characteristic of some of the limestones of this country. This odour was most perceptible in the specimens numbered 8, 9, 13, 19, 20.

In all the specimens in which the quantity of material was sufficient, the alkaline vapours which accompanied the moisture evolved were readily recognized.

The portion of the sample taken for analysis, after being treated with hydrochloric acid, yielded in every case a residue of a whitish-grey colour, Nos. 10, 11, and 12 being nearly white.

No. 8.—Material at command 9·80 grains.

Loss on ignition	0·895
Soluble in acid	7·506
Insoluble in acid	1·399
	9·800

No. 9.—Material at command, 9·10 grains.

Loss on ignition	0·80
Soluble in acid	7·30
Insoluble in acid	1·00
	9·10

No. 10.—Material at command, 19·60 grains.

Loss on ignition	1·30
Soluble in acid	16·10
Insoluble in acid	2·20
	19·60

No. 11.—Material at command, 24 grains.

Loss on ignition	1·10
Soluble in acid	21·80
Insoluble in acid	1·10
	24·00

No. 12.—Material at command, 12·0 grains.

Loss on ignition	0·50
Soluble in acid	10·50
Insoluble in acid	1·00
	12·00

No. 14.—Material at command, 27·80 grains.

Loss on ignition	1·90
Soluble in acid	23·20
Insoluble in acid	2·70
	27·80

APPEN

Table showing the relative frequency of the occurrence of the principal
Trawling was carried to a

	Station VII. 2125 fms.	Station 5. 2740 fms.	Station 9. 3150 fms.	Station 20. 2975 fms.	Station 29. 2700 fms.	Station 40. 2675 fms.	Station 54. 2650 fms.	Station 61. 2850 fms.	Station 63. 2750 fms.	Station 64. 2750 fms.	Station 68. 2175 fms.	Station 69. 2200 fms.	Station 89. 2400 fms.	Station 101. 2500 fms.	Station 104. 2400 fms.	Station 131. 2275 fms.	Station 134. 2025 fms.	Station 137. 2550 fms.	Station 160. 2600 fms.	Station 165. 2600 fms.	Station 181. 2440 fms.	Station 198. 2150 fms.
Pisces	...	...	...	...	...	*	...	...	*	...	...	...	*	*	*	...	...	*	...	...	*	*
Cephalopoda	...	...	...	...	...	...	...	...	...	...	...	...	...	...	...	...	...	...	...	...	...	...
Gastropoda	...	...	...	...	...	...	...	...	...	...	...	...	*	...	*	...	...	...	*	...	...	...
Lamellibranchiata	...	*	...	...	...	...	...	*	...	...	...	...	*	...	...	*	*	...	...	*
Brachiopoda	...	...	...	...	...	...	...	...	...	...	...	...	...	...	...	...	...	...	*	*	...	*
Tunicata	...	...	...	...	...	...	*	*	...	...	...	...	...	...	...	...	...	...	*	...	...	...
Pycnogonida	...	...	...	...	...	...	...	...	...	...	...	...	...	...	...	...	...	...	...	...	...	...
Decapoda	...	...	...	...	*	*	*	*	...	...	*	...	...	*	*	...	...	*	...	*	*	*
Schizopoda	...	...	...	...	...	...	...	...	...	...	...	...	*	...	...	...	...	...	...	...	...	...
Edriophthalmata	...	...	...	...	...	...	...	...	...	...	...	*	...	...	*	...	...	...	...	...	...	*
Cirripedia	...	...	...	...	...	...	...	...	...	...	...	...	*	...	*	...	...	...	*	...	...	*
Annelida	...	...	...	*	...	*	*	*	*	...	...	...	*	*	...	...	...	*	*	...	...
Gephyrea	...	...	...	...	...	...	...	...	...	*	...	...	...	*	...	...	...	...	...	...	...	...
Bryozoa	...	*	...	...	...	...	...	...	...	*	*	...	*	*	*	...	...	...	*	*	...	...
Holothuridea	...	...	...	...	...	...	*	...	...	...	...	*	*	*	...	...	...	*	*	...	...
Echinoidea	...	...	...	...	...	*	...	...	...	...	...	...	...	...	*	...	*	...	...	...
Ophiuridea	...	...	...	...	...	*	*	...	*	...	...	*	...	*	...	...	...	*	...	*	*
Asteridea	*	...	...	...	...	...	...	...	...	...	...	...	...	*	...	...	*	...	*	...	...
Crinoidea	...	...	...	...	...	...	...	...	...	...	...	...	...	...	...	...	...	*	...	...	...
Hydromedusæ	...	...	...	...	...	...	*	...	...	...	...	...	...	...	...	...	...	...	...	...	*
Zoantharia	...	...	...	...	...	...	...	*	...	*	...	...	...	...	*	...	...	*	...	*	...
Alcyonaria	*	...	...	...	...	...	...	...	...	*	*	...	...	...	...	...	...	*	*	...
Porifera	...	...	...	...	...	...	*	...	...	*	...	...	*	...	...	...	...	*	*	*	*

DIX B.

groups of Marine Animals at Fifty-two Stations at which Dredging or depth greater than 2,000 fathoms.

	Station 206. 2170 fms.	Station 215. 2500 fms.	Station 216. 2000 fms.	Station 223. 2925 fms.	Station 226. 2300 fms.	Station 230. 2425 fms.	Station 241. 2300 fms.	Station 244. 2900 fms.	Station 246. 2900 fms.	Station 252. 2740 fms.	Station 253. 3125 fms.	Station 264. 3000 fms.	Station 271. 2425 fms.	Station 272. 2600 fms.	Station 274. 2750 fms.	Station 276. 2350 fms.	Station 281. 2385 fms.	Station 285. 2375 fms.	Station 286. 2335 fms.	Station 289. 2550 fms.	Station 291. 2250 fms.	Station 293. 2025 fms.	Station 298. 2225 fms.	Station 299. 2160 fms.	Station 318. 2040 fms.	Station 325. 2650 fms.	Station 332. 2200 fms.	Station 333. 2025 fms.	Station 346. 2350 fms.	Station 348. 2450 fms.

APPENDIX C.

Table showing the amount of Carbonic Acid contained in Sea-Water at various Stations in the Atlantic.

Date. 1873.	Latitude.	Longitude.	Depth of Sample in Fathoms.	Temperature at Depth.	Specific Gravity of water at 15°·56 C.; water at 4° C. = 1.	Grammes of CO₂ in one litre of water.
Feb. 28	23° 10′ N.	38°42′ W.	2720 Bottom.	2°· 0C.	1·02747	0·04
Mar. 26	19 41	65 7	3875 Bottom.	*—	1·02637	0·057
27	21 26	65 16	Surface.	25 · 2	1·02703	0·046
28	22 49	65 19	2960 Bottom.	1 ·50	1·02597	0·053
29	24 39	65 25	2850 Bottom.	1 ·67	1·02606	0·052
31	27 49	64 59	Surface.	—	1·02736	0·048
May 26	36 30	63 40	2650 Bottom.	1 · 8	1·02690	0·064
27	34 50	63 59	Surface.	21 · 7	1·02711	0·045
June 14	32 54	63 22	Surface.	23 · 3	1·02716	0·0415
16	,, ,,	,, ,,	2360 Bottom.	1 · 7	1·02650	0·0472
	34 28	58 56	2575 Bottom.	1 · 5	1·02701	0·0500
23	37 54	41 44	Surface.	21 · 1	1·02690	0·0529
24	38 3	39 19	2175 Bottom.	—	1·02607	0·0536
27	38 18	34 48	1675 Bottom.	2 · 3	1·02660	0·0592
30	38 30	31 14	1000 Bottom.	3 · 7	1·02683	0·0446
Aug. 16	7 1	15 55	Surface.	26 · 1	1·02615	0·0432
18	6 11	15 57	Surface.	26 · 0	1·02637	0·0382
19	5 48	14 20	Surface.	26 · 2	1·02635	0·0455
20	4 29	13 52	Surface.	26 · 2	1·02622	0·0430
21	3 8	14 49	300	5 · 3	1·02610	0·0536
25	1 47	24 26	Surface.	26 · 0	1·02618	0·0426
26	1 47	24 26	50	—	1·02630	0·0533
Sept. 27	14 51 S.	37 1	Surface.	25 · 3	1·02770	0·0330
30	20 13	35 19	100	17 · 3	1·02736	0·0360
Oct. 1	22 15	35 37	Surface.	22 · 8	1·02744	0·0591
2	24 43	34 17	Surface.	21 · 0	1·02717	0·0418
3	26 15	32 56	2350 Bottom.	0 · 8	1·02706	0·0491
4	27 43	31 3	Surface.	19 · 4	1·02702	0·0432
6	29 35	28 9	1000	2 · 5	1·02572	0·0556

* On this occasion two thermometers were crushed by the extreme pressure.

INDEX.

INDEX.

A.

Abel, Prof., F.R.S., report on samples of soil from Bermudas, i. 348.
Acanthometrina, i. 233.
Aceste bellidifera, i. 376.
Açores, The, ii. 18.
Aërope rostrata, i. 380.
Adansonia gigantea, ii. 75.
'Æolian' rocks, Bermudas, i. 307, 313, 347.
African current or Guinea current, ii. 79.
Agulhas current, ii. 311.
Albatross, ii. 146, 159, 164, 183.
Alciope, i. 177.
Aldrich, Lieut. Pelham, R.N., appointed to the 'Challenger,' i. 10. ; ii. 92.
Algesiras, i. 127.
Alima, i. 178.
Aloes in flower, i. 126.
Altingia excelsa, San Miguel, ii. 27.
American deep-sea expeditions, i. 8.
Ammocharidæ, i. 201.
Amphinomidæ, i. 176.
Annelid, tube-building, living at the sea-bottom, i. 201.
Annelids in the Atlantic, ii. 348.
Antennarius marmoratus, i. 194 ; its nests of gulf-weed, ii. 10.
Aqueduct at Algesiras, i. 127.
Arca, living on the sea-bottom, i. 179, 246.
Arrowroot, its cultivation in Bermudas, i. 340.
Ascension, Island of, ii. 257, 267 ; George Town, Green Mountain, 258 ; government, 259, 262 ; botany, 261 ; climate, 262 ; 'Wide-awake Fair,' 264 ; birds, *ibid.*
Astacidæ, Willemöesia leptodactyla, i. 187, 258.
Astacus pellucidus, i. 190.
Astacus zaleucus, i. 259.
Astronomical Observatory, Lisbon, i. 116.
Atlantic, The : General conclusions from the 'Challenger' Expedition, ii. 287—368 ; contour of the bed, 288—291 ; nature of the bottom, 291—300 ; dis-tribution of temperature, surface and submarine currents, 300—328 ; density of sea-water, 354—363 ; amount of carbonic acid and oxygen in sea-water, 363—368.
Atlanta peronii, i. 123.
Avicula, i. 195.
Avocada pear, i. 343.

B.

Bailey, Prof., globigerina, i. 208.
Baillie, C. W., Navigating Lieut., his sounding-machine, i. 47.
Balanoglossus, ii. 82.
Ball's dredge, i. 49.
Balsam-bog,' Falkland Islands, ii. 210.
Baobab-tree, Cape Verde Islands, ii. 75.
Barnacles, ii. 4.
Barometer, aneroid, by Messrs. Elliott, used in the Expedition, i. 157.
Barometrical observations taken during the expedition, explanatory diagrams, i. 157.
Barometrical pressure, its relation to latitude, i. 79.
Basalt rocks, Fernando Neronha, ii. 117, 119 ; Tristan d'Acunha, 164.
Bathycrinus aldrichianus, ii. 93.
Beach marks, i. 76.
Belem : Castle, i. 108 ; monastery of Santa Maria, porch, quadrangle, and cloisters, 110, 112, 113.
Bethell, Lieut. George R., R.N., ap-pointed to the 'Challenger,' i. 10, 170, 242.
Bermudas, history and description of, i. 299.
Bermudas arrowroot, i. 340.
Bignonia at Madeira, i. 152.
Birds of Bermudas, i. 299, 324, 345 ; Brazil, ii. 140 ; Falkland Islands, 205 ; Island of Ascension, 264 ; San Miguel, 41 ; Tristan d'Acunha, 165, 177, 178.
Black coral, ii. 63.
Blind crustaceans, i. 255.

Papaw-trees, Bermudas, i. 344.
Peat of Falkland Islands, ii. 212.
Pelagic foraminifera, ii. 293, 343.
Penal servitude at Fernando Noronha, ii. 112, 115.
Penguins on Inaccessible Island, ii. 167, 179—183; on Nightingale Island, 186, 195.
Pentacrinus maclearanus, ii. 123, 124.
Persea gratissima, i. 343.
Petrels, i. 199; at Tristan d'Acunha, ii. 177, 265.
Philippine Islands, specimens of *Euplectella* from the, i. 136.
Phonolite rocks at Fernando Noronha, ii. 117.
Phormosoma uranus, i. 146; *hoplacantha*, 148.
Phosphorescence of the sea, i. 186, 190; ii. 85.
Phosphorescent animals, i. 178.
Phosphorescence of *Gorgonoid*, i. 119; of *umbellularia grœnlandica*, i. 151.
Photography on board the 'Challenger,' i. 46; Engravings from Photographs:— Belem Castle, i. 112; Cloister of Santa Maria, Belem, 113. Bermudas: Group of palms, 'Æolian' Rocks, Land-glaciers, Convolvulus Cave, Calcareous concretions, Cedar Avenue, swamp-vegetation, Papaw-trees, 301, 309, 310, 312, 313, 319, 325, 330, 331, 332, 333, 334, 338, 341, 344, 347; natives of San Vicente, i. 404; Açores; garden trees at San Miguel, ii. 27, 29, 32; orange groves, 35; St. Paul's Rocks, 104, 107; breeding place of the noddy, 104; Tristan d'Acunha: 'Edinburgh' settlement, 159; Cyclopean architecture, 162; Inaccessible Island, waterfall, 166; group of rock-hoppers, 179; group of penguins, 195; irrigation, Porto Prayo, 271.
Phylica arborea, at Tristan d'Acunha, ii. 159, 177.
Physalia, i. 120.
Pico, Island of, Açores, ii. 19, 24.
Plagusiæ, ii. 90.
Plants in the Atlantic, ii. 338.
Platform Island, Fernando Noronha group, ii. 119.
Poison ivy' of Bermudas, i. 329.
Polycystina, i. 233.
Polyzoa in the Atlantic, ii. 348.
Ponta Delgada, San Miguel, ii. 19, 26, 35.
Porcellanaster ceruleus, i. 379.
'Porcupine,' H.M.S., sounding expedition, i. 6; temperature soundings, 240, 246.
Port Louis, Falkland Island, ii. 206.
Porto Prayo, Cape Verde Islands, ii. 74, 77; mode of irrigation, 271.
Portuguese men-of-war,' i. 120.
Potato cultivation in Bermudas, i. 342.

Pourtales, Count, on globigerina and orbulina, i. 208, 216; on 'deep-sea corals,' 265, 269.
'Pride of India' (*Melia azedarach*), Bermudas, i. 322.
Protective resemblance in the gulf-weed fauna, ii. 10.
Protozoa, i. 231.
Psolus ephippiffer, ii. 220, 222
Pteropoda, i. 119, 123.
Pterotrachea, i. 121.
Pulvinulina, i. 187, 206—219, 217.
Pumice fragments on the bed of the Atlantic, i. 229, 239; ii. 296.
Pycnogonida in the Atlantic, ii. 349.
Pyrocystis fusiformis, ii. 89.
Pyrocystis noctiluca, ii. 88.
Pyrosoma, phosphorescence of, i 186; ii. 85.

R.

Radiolaria, i. 179, 181, 186, 206, 208, 230, 231; ii. 340.
Rainfall at the Island of Ascension, ii. 260, 297.
Rain water in universal use in Bermudas, i. 297.
Rat Island, Fernando Noronha group, ii. 118.
Red-clay of the bed of the Atlantic, its nature and origin, i. 226, 182, 187, 193, 196, 200, 223, 225, 279, 285, 315; ii. 296, 299.
Red-clay and globigerina-ooze, tabular view of their proportions, i. 226.
Red coral fishery, Cape Verde Islands, ii. 76, 77.
'Red earth' of Bermudas, i. 315, 316, 351.
Religious ceremony in San Miguel, ii. 48.
Rhabdoliths, i. 206, 220, 228.
Rhabdospheres, specimens from the surface, 500 and 2,000 times the natural size, i. 221, 222.
Rhizocrinus, ii. 99.
Rhizopods, i. 228.
Rhus toxicodendron, of Bermudas, i. 329.
Ribiera Grande, San Miguel, ii. 35.
Richards, Admiral, Sir George, F.R.S., his aid to the Expedition, i. 9, 69.
'Rock-hoppers,' on Inaccessible Island, ii. 167, 179.
Rotalia, i. 208.
Royal Society, proposal for a circumnavigating expedition, i. 8.

S.

St. Elmo's fires, i. 291.
St. George's Island, Bermudas, i. 297, 337.
St. Michael's Mount, Fernando Noronha, ii. 117.

THE END.

LONDON : R. CLAY, SONS, AND TAYLOR, PRINTERS.